Revise
A2

OCR
Mathematics

Peter Sherran & Janet Crawshaw

Contents

Chapter 3 Mechanics 2

Chapter 4 Decision Mathematics 2

Specification information

OCR Mathematics

MODULE	SPECIFICATION TOPIC	CHAPTER REFERENCE	STUDIED IN CLASS	REVISED	PRACTICE QUESTIONS
Core 3 (C3)	Algebra and functions	1.2, 1.6			
	Trigonometry	1.4			
	Differentiation and integration	1.5, 1.6			
	Numerical methods	1.7			
Core 4 (C4)	Algebra and graphs	1.1, 1.3			
	Differentiation and integration	1.5, 1.6			
	First order differential equations	1.6			
	Vectors	1.8			
Probability and Statistics 2 (S2)	Continuous random variables	2.1			
	The normal distribution	2.1, 2.2			
	The Poisson distribution	2.2			
	Sampling and hypothesis tests	2.3, 2.4			
Mechanics 2 (M2)	Centre of mass	3.3			
	Equilibrium of a rigid body	3.2			
	Motion of a projectile	3.1			
	Uniform motion in a circle	3.5			
	Coefficient of restitution; impulse	3.4			
	Energy, work and power	3.6			
Decision 2 (D2)	Game theory	4.1			
	Flows in a network	4.2			
	Matching and allocation problems	4.4			
	Critical path analysis	4.3			
	Dynamic programming	4.5			

Examination analysis

The Advanced Level Mathematics GCE consists of AS (50%) and A2 (50%). AS consists of C1 + C2 + one of M1, S1, D1. A2 consists of C3 + C4 + one of M1, S1, D1, M2, S2, D2 (following any dependency rules).

C3	*A2*	*Scientific/graphics calculator*	*1 hr 30 min exam*	*$16\frac{2}{3}$% of A Level*
C4	*A2*	*Scientific/graphics calculator*	*1 hr 30 min exam*	*$16\frac{2}{3}$% of A Level*
M1/M2	*AS/A2*	*Scientific/graphics calculator*	*1 hr 30 min exam*	*$16\frac{2}{3}$% of A Level*
S1/S2	*AS/A2*	*Scientific/graphics calculator*	*1 hr 30 min exam*	*$16\frac{2}{3}$% of A Level*
D1/D2	*AS/A2*	*Scientific/graphics calculator*	*1 hr 30 min exam*	*$16\frac{2}{3}$% of A Level*

Your AS/A2 Level Mathematics course

AS and A2

The OCR Mathematics A Level course is in two parts, with three separate modules in each part. Students first study the AS (Advanced Subsidiary) course. Some will then go on to study the second part of the A Level course, called A2. Advanced Subsidiary is assessed at the standard expected halfway through an A Level course: i.e., between GCSE and Advanced GCE. This means that the AS and A2 courses are designed so that difficulty steadily increases:

- AS Mathematics builds from GCSE Mathematics
- A2 Mathematics builds from AS Mathematics.

How will you be tested?

Assessment units

For AS Mathematics, you will be tested by three assessment units. For the full A Level in Mathematics, you will take a further three units. AS Mathematics forms 50% of the assessment weighting and A2 Mathematics forms the other 50% for the full A Level.

Most units can be taken in either January or June, but some A2 units must be taken at the end of the course. There is a lot of flexibility about when exams can be taken and the diagram below shows just some of the ways that the assessment units may be taken for AS and A Level Mathematics.

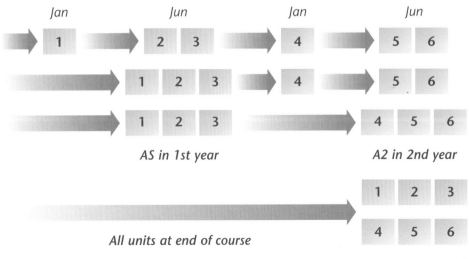

If you are disappointed with a module result, you can resit the module. There is no restriction on the number of times a module may be attempted. The best available results for each module will count towards the final grade.

A2 progression

After having studied AS Mathematics, to continue studying Mathematics to A Level you will need to take three further units of Mathematics, at least two of which are at A2 standard.

Synoptic assessment

Synoptic assessment in Mathematics addresses candidates' understanding of the connections between different elements of the subject. It involves the explicit drawing together of knowledge, understanding and skills learned in different parts of the A Level course, through using and applying methods developed earlier to solving problems. Making and understanding connections in this way is intrinsic to learning mathematics.

For A2 modules you will be asked to demonstrate knowledge of modules studied at AS Level. The contents of C1 and C2 are assumed for C3 and the contents of C1, C2 and C3 are assumed for C4. Similarly in the application modules, a knowledge of S1 is assumed for S2, a knowledge of M1 is assumed for M2 and a knowledge of D1 is assumed for D2.

Key skills

It is important that you develop your key skills throughout your AS and A2 courses. These are important skills that you need whatever you do beyond AS and A Levels. To gain the key skills qualification, which is equivalent to an AS Level, you will need to collect evidence together in a 'portfolio' to show that you have attained Level 3 in Communication, Application of number and Information technology. You will also need to take a formal test in each key skill. You will have many opportunities during AS and A2 Mathematics to develop your key skills.

It is a worthwhile qualification, as it demonstrates your ability to put your ideas across to other people, collect data and use up-to-date technology in your work.

What skills will I need?

For A Level Mathematics (AS and A2), you will be tested by assessment objectives: these are the skills and abilities that you should have acquired by studying the course. The assessment objectives for A Level Mathematics are shown below.

Candidates should be able to:

- recall, select and use their knowledge of mathematical facts, concepts and techniques in a variety of contexts

- construct rigorous mathematical arguments and proofs through use of precise statements, logical deduction and inference and by the manipulation of mathematical expressions, including the construction of extended arguments for handling substantial problems presented in unstructured form

- recall, select and use their knowledge of standard mathematical models to represent situations in the real world; recognise and understand given representations involving standard models; present and interpret results from such models in terms of the original situation, including discussion of the assumptions made and refinement of such models

- comprehend translations of common realistic contexts into mathematics; use the results of calculations to make predictions, or comment on the context; and, where appropriate, read critically and comprehend longer mathematical arguments or examples of applications

- use contemporary calculator technology and other permitted resources (such as formulae booklets or statistical tables*) accurately and efficiently; understand when not to use such technology, and its limitations; give answers to appropriate accuracy.

* You can find a copy of the formulae booklet and statistical tables in the A Level mathematics section of the OCR website, www.ocr.org.uk

Exam technique

What are the examiners looking for?

Examiners use certain words in their instructions to let you know what they are expecting in your answer. Make sure that you know what they mean so that you can give the right response.

Write down, state

You can write your answer without having to show how it was obtained. There is nothing to prevent you doing some working if it helps you, but if you are doing a lot then you might have missed the point.

Calculate, find, determine, show, solve

Make sure that you show enough working to justify the final answer or conclusion. Marks will be available for showing a correct method.

Deduce, hence

This means that you are expected to use the given result to establish something new. You must show all of the steps in your working.

Draw

This is used to tell you to plot an accurate graph using graph paper. Take note of any instructions about the scale that must be used. You may need to read values from your graph.

Sketch

If the instruction is to sketch a graph then you don't need to plot the points but you will be expected to show its general shape and its relationship with the axes. Indicate the positions of any turning points and take particular care with any asymptotes.

Find the exact value

This instruction is usually given when the final answer involves an irrational value such as a logarithm, e, π or a surd. You will need to demonstrate that you can manipulate these quantities so don't just key everything into your calculator or you will lose marks.

If a question requires the final answer to be given to a specific level of accuracy then make sure that you do this or you might needlessly lose marks.

Some dos and don'ts

Dos

Do read the question

- Make sure that you are clear about what you are expected to do. Look for some structure in the question that may help you take the right approach.
- Read the question *again* after you have answered it as a quick check that your answer is in the expected form.

Do use diagrams

- In some questions, particularly in mechanics, a clearly labelled diagram is essential. Use a diagram whenever it may help you understand or represent the problem that you are trying to solve.

Do take care with notation

- Write clearly and use the notation accurately. Use brackets when they are required.
- Even if your final answer is wrong, you may earn some marks for a correct expression in your working.

Do learn relevant formulae

- Each module specification contains a list of formulae that *will not be given* in the examination. Make sure that you learn these formulae and practise using them.
- Some formulae *will be given* in the Examination Formulae Booklet. Make sure you know what they are and where to find them. This will help you refer to them quickly in the examination so that you don't waste time.

Do avoid silly answers

- Check that your final answer is sensible within the context of the question.

Do make good use of time

- Choose the order in which you answer the questions carefully. Do the ones that are easiest for you first.
- Set yourself a time limit for a question depending on the number of marks available.
- Be prepared to leave a difficult part of a question and return to it later if there is time.
- Towards the end of the exam make sure that you pick up all of the easy marks in any questions that you haven't got time to answer fully.

Don'ts

Don't work with rounded values

- There may be several stages in a solution that produce numerical values. Rounding errors from earlier stages may distort your final answer. One way to avoid this is to make use of your calculator memories to store values that you will need again.

Don't cross out work that may be partly correct

- It's tempting to cross out something that hasn't worked out as it should. Avoid this unless you have time to replace it with something better.

Don't write out the question

- This wastes time. The marks are for your solution!

What grade do you want?

Everyone should be able to improve their grades but you will only manage this with a lot of hard work and determination. The details given below describe a level of performance typical of candidates achieving grades A, C or E. You should find it useful to read and compare the expectations for the different levels and to give some thought to the areas where you need to improve most.

Grade A candidates

- Recall or recognise almost all the mathematical facts, concepts and techniques that are needed, and select appropriate ones to use in a variety of contexts.
- Manipulate mathematical expressions and use graphs, sketches and diagrams, all with high accuracy and skill.
- Use mathematical language correctly and proceed logically and rigorously through extended arguments or proofs.
- When confronted with unstructured problems they can often devise and implement an effective solution strategy.
- If errors are made in their calculations or logic, these are sometimes noticed and corrected.
- Recall or recognise almost all the standard models that are needed, and select appropriate ones to represent a wide variety of situations in the real world.
- Correctly refer results from calculations using the model to the original situation; they give sensible interpretations of their results in the context of the original realistic situation.
- Make intelligent comments on the modelling assumptions and possible refinements to the model.
- Comprehend or understand the meaning of almost all translations into mathematics of common realistic contexts.
- Correctly refer the results of calculations back to given context and usually make sensible comments or predictions.
- Can distil the essential mathematical information from extended pieces of prose having mathematical content.
- Comment meaningfully on the mathematical information.
- Make appropriate and efficient use of contemporary calculator technology and other permitted resources, and are aware of any limitations to their use.
- Present results to an appropriate degree of accuracy.

Grade C candidates

- Recall or recognise most of the mathematical facts, concepts and techniques that are needed, and usually select appropriate ones to use in a variety of contexts.
- Manipulate mathematical expressions and use graphs, sketches and diagrams, all with a reasonable level of accuracy and skill.
- Use mathematical language with some skill and sometimes proceed logically through extended arguments or proofs.
- When confronted with unstructured problems they sometimes devise and implement an effective and efficient solution strategy.
- Occasionally notice and correct errors in their calculations.
- Recall or recognise most of the standard models that are needed and usually select appropriate ones to represent a variety of situations in the real world.

- Often correctly refer results from calculations using the model to the original situation; they sometimes give sensible interpretations of their results in context of the original realistic situation.
- Sometimes make intelligent comments on the modelling assumptions and possible refinements to the model.
- Comprehend or understand the meaning of most translations into mathematics of common realistic contexts.
- Often correctly refer the results of calculations back to the given context and sometimes make sensible comments or predictions.
- Distil much of the essential mathematical information from extended pieces of prose having mathematical content.
- Give some useful comments on this mathematical information.
- Usually make appropriate and effective use of contemporary calculator technology and other permitted resources, and are sometimes aware of any limitations to their use.
- Usually present results to an appropriate degree of accuracy.

Grade E candidates

- Recall or recognise some of the mathematical facts, concepts and techniques that are needed, and sometimes select appropriate ones to use in some contexts.
- Manipulate mathematical expressions and use graphs, sketches and diagrams, all with some accuracy and skill.
- Sometimes use mathematical language correctly and occasionally proceed logically through extended arguments or proofs.
- Recall or recognise some of the standard models that are needed and sometimes select appropriate ones to represent a variety of situations in the real world.
- Sometimes correctly refer results from calculations using the model to the original situation; they try to interpret their results in the context of the original realistic situation.
- Sometimes comprehend or understand the meaning of translations into mathematics of common realistic contexts.
- Sometimes correctly refer the results of calculations back to the given context and attempt to give comments or predictions.
- Distil some of the essential mathematical information from extended pieces of prose having mathematical content; they attempt to comment on this mathematical information.
- Candidates often make appropriate and efficient use of contemporary calculator technology and other permitted resources.
- Often present results to an appropriate degree of accuracy.

The table below shows how your uniform standardised mark is translated.

UMS	80%	70%	60%	50%	40%
grade	A	B	C	D	E

To achieve an A* grade in Mathematics, you need to achieve a grade A overall (an average of 80 or more on uniform mark scale) for the whole A Level qualification and an average of 90 or more on the uniform mark scale in Core 3 and Core 4. It is awarded for A Level qualification only and not for the AS qualification or individual units.

Four steps to successful revision

Step 1: Understand

- Study the topic to be learned slowly. Make sure you understand the logic or important concepts.
- Mark up the text if necessary – underline, highlight and make notes.
- Re-read each paragraph slowly.

GO TO STEP 2

Step 2: Summarise

- Now make your own revision note summary:
  What is the main idea, theme or concept to be learned?
  What are the main points? How does the logic develop?
  Ask questions: Why? How? What next?
- Use bullet points, mind maps, patterned notes.
- Link ideas with mnemonics, mind maps, crazy stories.
- Note the title and date of the revision notes
  (e.g. Mathematics: Differentiation, 3rd March).
- Organise your notes carefully and keep them in a file.

This is now in **short-term memory**. You will forget 80% of it if you do not go to Step 3.
GO TO STEP 3, but first take a 10 minute break.

Step 3: Memorise

- Take 25 minute learning 'bites' with 5 minute breaks.
- After each 5 minute break test yourself:
  Cover the original revision note summary
  Write down the main points
  Speak out loud (record yourself)
  Tell someone else
  Repeat many times.

The material is well on its way to **long-term memory**.
You will forget 40% if you do not do step 4. **GO TO STEP 4**

Step 4: Track/Review

- Create a Revision Diary (one A4 page per day).
- Make a revision plan for the topic, e.g. 1 day later, 1 week later, 1 month later.
- Record your revision in your Revision Diary, e.g.
  Mathematics: Differentiation, 3rd March 25 minutes
  Mathematics: Differentiation, 5th March 15 minutes
  Mathematics: Differentiation, 3rd April 15 minutes
  ... and then at monthly intervals.

Chapter 1

Core 3 and Core 4 (Pure Mathematics)

The following topics are covered in this chapter:

- Algebra and series
- Functions including exponential and logarithmic
- Coordinate geometry
- Trigonometry

- Differentiation
- Integration
- Numerical methods
- Vectors

1.1 Algebra and series

After studying this section you should be able to:

- simplify rational expressions
- express rational expressions in partial fractions
- use the binomial expansion $(1 + x)^n$ for any rational n

LEARNING SUMMARY

Rational expressions

Core 4

A **rational expression** is of the form $\dfrac{f(x)}{g(x)}$, where $f(x)$ and $g(x)$ are polynomials in x.

To simplify a rational expression, factorise both $f(x)$ and $g(x)$ as far as possible, then cancel any factors that appear in both the numerator and denominator.

Examples

(a) $\dfrac{2x^2 + 6x}{2x^2 + 7x + 3} = \dfrac{2x(x + 3)}{(2x + 1)(x + 3)}$

$= \dfrac{2x}{2x + 1}$

> This cannot be cancelled any further.

Factorise $x^2 - 4$ using difference between two squares.

(b) $\dfrac{3x^2 - 12}{2 - x} = \dfrac{3(x^2 - 4)}{2 - x}$

$x - 2 = -(2 - x).$

$= \dfrac{3(x - 2)(x + 2)}{(2 - x)}$

$= -3(x + 2)$

When adding and subtracting algebra fractions, always use the lowest common denominator, for **example**,

First factorise the denominators.

The lowest common multiple of the denominators is $(x + 3)(x - 4)(x - 3)$.

$\dfrac{x - 2}{x^2 - x - 12} + \dfrac{4}{x^2 - 9} = \dfrac{x - 2}{(x + 3)(x - 4)} + \dfrac{4}{(x + 3)(x - 3)}$

$= \dfrac{(x - 3)(x - 2) + 4(x - 4)}{(x + 3)(x - 4)(x - 3)}$

$= \dfrac{x^2 - 5x + 6 + 4x - 16}{(x + 3)(x - 4)(x - 3)}$

$= \dfrac{x^2 - x - 10}{(x + 3)(x - 4)(x - 3)}$

Key points from AS

- **Remainder theorem**
  page 38
- **Algebraic division**
  page 37

Partial fractions

Core 4

It is sometimes possible to decompose a rational function into **partial fractions**. This format is often useful when differentiating, integrating or expanding functions as a series. Look out for the following types of partial fractions.

Denominator contains linear factors only

$$\frac{x+7}{(x-2)(x+1)} \equiv \frac{A}{x-2} + \frac{B}{x+1} = \frac{A(x+1)+B(x-2)}{(x-2)(x+1)}$$

The symbol $\equiv$ indicates that this is an identity. It is true for all values of x.

$$\Rightarrow x+7 \equiv A(x+1) + B(x-2)$$

These substitutions are chosen so that each factor in turn becomes zero.

Let $x = -1$, then $6 = -3B \Rightarrow B = -2$
Let $x = 2$, then $9 = 3A \Rightarrow A = 3$

$$\therefore \frac{x+7}{(x-2)(x+1)} \equiv \frac{3}{x-2} - \frac{2}{x+1}$$

Denominator contains repeated linear factors

Do not use $(x+3)(x+1)(x+1)^2$ as the common denominator as it is not the lowest common multiple.

$$\frac{6x^2+15x+7}{(x+3)(x+1)^2} \equiv \frac{A}{x+3} + \frac{B}{x+1} + \frac{C}{(x+1)^2}$$

$$\equiv \frac{A(x+1)^2 + B(x+3)(x+1) + C(x+3)}{(x+3)(x+1)^2}$$

$$\Rightarrow 6x^2 + 15x + 7 \equiv A(x+1)^2 + B(x+3)(x+1) + C(x+3)$$

Use a mixture of substitution and equating coefficients to find A, B and C.

Let $x = -1$, then $-2 = 2C \Rightarrow C = -1$
Let $x = -3$, then $16 = 4A \Rightarrow A = 4$
Equate x^2 terms: $6 = A + B$ and since $A = 4$, $B = 2$.

$$\therefore \frac{6x^2+15x+7}{(x+3)(x+1)^2} \equiv \frac{4}{x+3} + \frac{2}{x+1} - \frac{1}{(x+1)^2}$$

The degree of a polynomial is its highest power of x, for example $x^3 + 2x^2 - 3x + 4$ has degree 3.

Only **proper fractions** (when the numerator is of lower degree than the denominator) can be written in partial fraction form. For **improper fractions**, divide the denominator into the numerator first. This can be done by long division or by using identities.

Binomial expansion for rational n

Core 4

> When n is rational, $(1 + x)^n$ can be written as a series, using the binomial expansion
>
> $$(1+x)^n = 1 + nx + \frac{n(n-1)}{2!}x^2 + \frac{n(n-1)(n-2)}{3!}x^3 + \dots$$

KEY POINT

Key points from AS

- **Binomial expansion when n is a positive integer**
  Revise AS page 42

If n is a positive integer, the series terminates at the term in x^n.

For all other values of n, the series is infinite and converges provided that $|x| < 1$.

Example

(a) Expand $\sqrt{1 - 4x}$ as an ascending series in x, as far as the term in x^3, giving the set of values of x for which the series is valid.

(b) By substituting $x = 0.01$ into the series expansion of $\sqrt{1 - 4x}$, find $\sqrt{96}$ correct to 4 decimal places.

> $1 - 4x = 1 + (-4x)$, so use $(1 + x)^n$ with $n = \frac{1}{2}$ and substituting $(-4x)$ for x.

(a) $\sqrt{1 - 4x} = (1 - 4x)^{\frac{1}{2}}$

$$= 1 + \tfrac{1}{2}(-4x) + \frac{(\frac{1}{2})(-\frac{1}{2})}{2!}(-4x)^2 + \frac{(\frac{1}{2})(-\frac{1}{2})(-\frac{3}{2})}{3!}(-4x)^3 + \ldots$$

$$= 1 - 2x - 2x^2 - 4x^3 + \ldots$$

> You should also state the set of values of x for which the expansion is valid, even when it is not specifically requested in the question.

The series is valid provided that $|4x| < 1$, i.e. $|x| < \frac{1}{4}$.

(b) Substituting $x = 0.01$ into $\sqrt{1 - 4x}$ gives $\sqrt{0.96} = \sqrt{\dfrac{96}{100}} = \dfrac{\sqrt{96}}{10}$.

So $\sqrt{96} = 10(1 - 2(0.01) - 2(0.01)^2 - 4(0.01)^3 + \ldots)$

$$= 10(1 - 0.02 - 0.0002 - 0.000004 +)$$

$$= 9.7980 \ (4 \ \text{d.p.})$$

The function to be expanded must be in the form $(1 + \ldots)^n$ or $(1 - \ldots)^n$.
You may have to do some careful algebraic manipulation to get it in this form.

Example

Give the first four terms in the series expansion of $\dfrac{1}{(2 + x)}$.

$$\frac{1}{(2 + x)} = \frac{1}{2\left(1 + \dfrac{x}{2}\right)}$$

> Alternatively,
> $(2 + x)^{-1} = 2^{-1}\left(1 + \dfrac{x}{2}\right)^{-1}$.

$$= \frac{1}{2}\left(1 + \frac{x}{2}\right)^{-1}$$

$$= \frac{1}{2}\left(1 + (-1)\left(\frac{x}{2}\right) + \frac{(-1)(-2)}{2!}\left(\frac{x}{2}\right)^2 + \frac{(-1)(-2)(-3)}{3!}\left(\frac{x}{2}\right)^3 + \ldots\right)$$

$$= \frac{1}{2}\left(1 - \frac{x}{2} + \frac{x^2}{4} - \frac{x^3}{8} + \ldots\right)$$

$$= \frac{1}{2} - \frac{x}{4} + \frac{x^2}{8} - \frac{x^3}{16} + \ldots$$

Restriction on x: $\left|\dfrac{x}{2}\right| < 1$, i.e. $|x| < 2$.

Progress check

Core 4

1 Simplify:

(a) $\dfrac{2x^2 + 3x}{2x^2 + x - 3}$ (b) $\dfrac{9x - x^3}{3x^2 + x^3}$

Core 4

2 Express as a single fraction in its simplest form:

(a) $\dfrac{3(x - 4)}{(x + 2)(x - 1)} + \dfrac{2(x + 1)}{x - 1}$

(b) $\dfrac{x^2 - 3x + 2}{x^2 + 3x - 4} - \dfrac{1}{(x + 4)^2}$

Core 4

3 $\dfrac{4x^2 + x + 1}{(x + 2)(x - 3)(x + 1)} \equiv \dfrac{A}{x + 2} + \dfrac{B}{x - 3} + \dfrac{C}{x + 1}$

Find the values of A, B and C.

Core 4

4 Express in partial fractions:

$\dfrac{3x^2 + 4x + 5}{(x - 1)^2(x + 2)}$

Core 4

5 Write as a series in ascending powers of x as far as the term in x^3, stating the values of x for which the expansion is valid:

(a) $\dfrac{1}{1 + x}$ (b) $(1 - 3x)^{-2}$ (c) $\sqrt{4 + x}$

1 (a) $\dfrac{x-1}{x}$ (b) $\dfrac{3-x}{x}$

2 (a) $\dfrac{2x^2 + 9x - 8}{(x+2)(x-1)}$ (b) $\dfrac{x^2 + 2x - 9}{(x+4)^2}$

3 $A = 3$, $B = 2$, $C = -1$

4 $\dfrac{2}{x-1} + \dfrac{4}{(x-1)^2} + \dfrac{1}{x+2}$

5 (a) $1 - x + x^2 - x^3 + \cdots$ for $|x| < 1$
(b) $1 + 6x + 27x^2 + 108x^3 + \cdots$ for $|x| < \frac{1}{3}$
(c) $2 + \frac{1}{4}x - \frac{1}{64}x^2 + \frac{1}{512}x^3 + \cdots$ for $|x| < 4$

1.2 Functions including exponential and logarithmic

After studying this section you should be able to:

- *use function notation and understand domain and range*
- *understand composite functions and find the inverse of a one-one function*
- *sketch the graph of the inverse of a function from the graph of the function*
- *understand the properties of the modulus function*
- *use transformations to sketch graphs of related functions*
- *understand the definitions of e^x and ln x*

LEARNING SUMMARY

Functions

Core 3

A **function** may be thought of as a rule which takes each member x of a set and assigns, or **maps**, it to some value y known as its **image**.

x maps to y
y is the image of x.

$$x \longrightarrow \boxed{\text{Function}} \longrightarrow y$$

$f(x)$ is read as 'f of x'.

A letter such as f, g or h is often used to stand for a function. The function which squares a number and adds on 5, for example, can be written as $f(x) = x^2 + 5$. The same notation may also be used to show how a function affects particular values. For this function, $f(4) = 4^2 + 5 = 21$, $f(-10) = (-10)^2 + 5 = 105$ and so on.

An alternative notation for the same function is $f: x \mapsto x^2 + 5$.

> **KEY POINT**
>
> The set of values on which the function acts is called the **domain** and the corresponding set of image values is called the **range**.

The function $f(x) = x + 2$ with domain {3, 6, 10} is shown below.

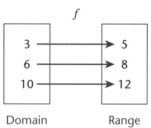

| Domain | | Range |

The symbol $\mathbb{R}$ is often used to stand for the complete set of numbers on the number line. This is the set of Real numbers.

The domain of a function may be an infinite set such as $\mathbb{R}$ but, in some cases, particular values must be omitted for the function to be valid.

For example, the function $f(x) = \dfrac{x+2}{x-3}$ cannot have 3 in its domain since division by zero is undefined.

Under a function, every member of the domain must have only one image. But, it is possible for different values in the domain to have the same image.

There are different types of function.

> **KEY POINT**
>
> If for each element y in the range there is *a unique* value of x such that $f(x) = y$, then f is a **one–one** function. If for any element y in the range there is more than one value of x satisfying $f(x) = y$, then f is **many–one**.

Examples

The function $f(x) = 2x$ with domain $\mathbb{R}$ is one–one.

The function $f(x) = x^2$ with domain $\mathbb{R}$ is many–one.

The function $f(x) = x^2$ with domain $x > 0$ is one–one.

The function $f(x) = \sin x$ with domain $0° \leqslant x \leqslant 360°$ is many–one.

The function $f(x) = \sin x$ with domain $-90° \leqslant x \leqslant 90°$ is one–one.

These examples show the importance of including the domain of the function in its definition.

Composition of functions

Core 3

The diagram shows how two functions f and g may be combined.

$$x \xrightarrow{\ \ f\ \ } f(x) \xrightarrow{\ \ g\ \ } g(f(x))$$

In this case, f is applied first to some value x giving $f(x)$. Then g is applied to the value $f(x)$ to give $g(f(x))$. This is usually written as $gf(x)$ and gf can be thought of as a new **composite function** defined from the functions f and g.

For example, if $f(x) = 3x$, $x \in \mathbb{R}$ and $g(x) = x + 2$, $x \in \mathbb{R}$ then
$$gf(x) = g(3x) = 3x + 2.$$

The order in which the functions are applied is important. The composite function fg is found by applying g first and then f.

In this case, $fg(x) = f(x + 2) = 3x + 6$.

Generally speaking, when two functions f and g are defined, the composite functions fg and gf will not be the same.

The inverse of a function

Core 3

The **inverse of a function** f is a function, usually written as f^{-1}, that *undoes* the effect of f. So the inverse of a function which adds 2 to every value, for example, will be a function that subtracts 2 from every value.

This can be written as $f(x) = x + 2$, $x \in \mathbb{R}$ and $f^{-1}(x) = x - 2$, $x \in \mathbb{R}$.

The domain of f^{-1} is given by the range of f.

Notice that $f^{-1}f(x) = f^{-1}(x + 2) = x$ and that $ff^{-1}(x) = f(x - 2) = x$.

A function can be either one–one or many–one, but only functions that are one–one can have an inverse. The reason is, that reversing a many–one function would give a mapping that is one–many, and this cannot be a function.

The diagram shows a many–one function. It does not have an inverse.

Every value in the domain of a function can have only one image. This is an important property of functions.

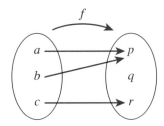

If you reverse this diagram then p would have more than one image.

Some important examples of restricting the domain of a function so that it will have an inverse are found in trigonometry. See page 28.

You can turn a many–one function into a one–one function by restricting its domain.

For example, the function $f(x) = x^2$, $x \in \mathbb{R}$ is many–one and so it does not have an inverse. However, the function $f(x) = x^2$, $x > 0$ is one–one and $f^{-1}(x) = \sqrt{x}$, $x > 0$.

Finding the inverse of a one–one function

Core 3

One way to find the inverse of a one–one function f is to write $y = f(x)$ and then rearrange this to make x the subject so that $x = f^{-1}(y)$. The inverse function is then usually defined in terms of x to give $f^{-1}(x)$. The domain of f^{-1} is the range of f.

You can turn $f^{-1}(y)$ into $f^{-1}(x)$ by replacing every y with an x.

Example   Find the inverse of the function $f(x) = \dfrac{x+2}{x-3}$, $x \neq 3$.

Start by writing $y = f(x)$.

Define $y = \dfrac{x+2}{x-3}$

Rearrange to find x.

then $y(x-3) = x+2$

Multiply out the brackets.

$xy - 3y = x + 2$

Collect the x terms on one side of the equation.

$xy - x = 3y + 2$

$x(y-1) = 3y + 2$

$x = \dfrac{3y+2}{y-1}$.

$x = f^{-1}(y)$ so this defines the inverse function in terms of y.

It is usual to express the inverse function in terms of x.

So $f^{-1}(x) = \dfrac{3x+2}{x-1}$, $x \neq 1$.

The denominator cannot be allowed to be zero so 1 is not in the domain of f^{-1}.

Here are the graphs of $y = \dfrac{x+2}{x-3}$ and its inverse $y = \dfrac{3x+2}{x-1}$

The graph of the inverse function is the reflection of the original graph in the line $y = x$. This is true for any inverse function but the same scale must be used on both axes or the effect is distorted.

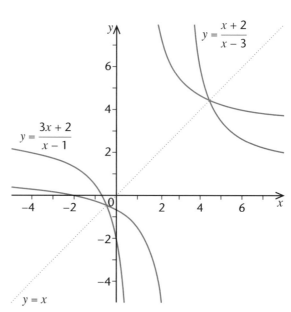

The modulus function

Core 3

The notation $|x|$ is used to stand for the modulus of x. This is defined as

$$|x| = \begin{cases} x & \text{when } x \geqslant 0 & \text{(when } x \text{ is positive, } |x| \text{ is just the same as } x\text{).} \\ -x & \text{when } x < 0 & \text{(when } x \text{ is negative, } |x| \text{ is the same as } -x\text{).} \end{cases}$$

The expression $|x-a|$ can be interpreted as the distance between the numbers x and a on the number line. In this way, the statement $|x-a| < b$ means that the distance between x and a is less than b.

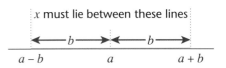
x must lie between these lines

It follows that $a - b < x < a + b$.

The graph of $y = |x|$

It follows that the graph of $y = |x|$ is the same as the graph of $y = x$ for positive values of x. But, when x is negative, the corresponding part of the graph of $y = x$ must be reflected in the x-axis to give the graph of $y = |x|$.

> $|x|$ is never negative so the graph of $y = |x|$ doesn't go below the x-axis anywhere.

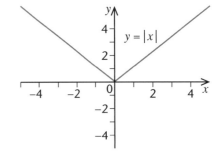

The graph of $y = |f(x)|$

The graph of $y = |f(x)|$ is the same as the graph of $y = f(x)$ for positive values of $f(x)$. But, when $f(x)$ is negative, the corresponding part of the graph of $y = f(x)$ must be reflected in the x-axis to give the graph of $y = |f(x)|$.

> The graph of $y = 1/x$ goes below the x-axis for negative values of x.

The diagram shows the graph of $y = \left|\dfrac{1}{x}\right|$.

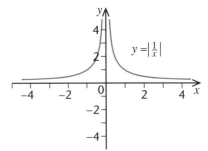

Transforming graphs

Core 3

The graph of some new function can often be obtained from the graph of a known function by applying a transformation. A summary of the standard transformations is given in the table.

> You may need to apply a combination of transformations in some cases.

Known function	New function	Transformation
$y = f(x)$	$y = f(x) + a$	Translation through a units parallel to y-axis.
	$y = f(x - a)$	Translation through a units parallel to x-axis.
	$y = af(x)$	One-way stretch with scale factor a parallel to the y-axis.
	$y = f(ax)$	One-way stretch with scale factor $\frac{1}{a}$ parallel to the x-axis.

21

Examples

$y = af(x)$ does not move points on the x-axis.

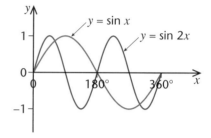

Sometimes you may have to perform more than one transformation.

Example

Key points from AS

- **Circles** *Revise AS page 25*

Describe how to transform the circle $x^2 + y^2 = 25$, with centre (0, 0) and radius 5, to the circle with centre (3, −8) and radius 5. Write down the new equation.

Translate 3 units parallel to the x-axis (to the right) and −8 units parallel to the y-axis (down).

$$x^2 + y^2 = 25 \rightarrow (x - 3)^2 + (y + 8)^2 = 25$$

The exponential function e^x

Core 3

Exponential functions are often used to represent, or model, patterns of:
- growth when $a > 1$
- decay when $a < 1$.

Key points from AS

- **Exponential functions** *Revise AS page 36*

All scientific calculators will display values of e^x and 10^x. To display the value of e, for example, find e^1.

An **exponential function** is one where the variable is a power or exponent. For example, any function of the form $f(x) = a^x$, where a is a constant, is an exponential function.

The diagram shows some graphs of exponential functions for different values of a.

All of the graphs pass through the point (0, 1) and each one has a different gradient at this point. The value of the gradient depends on a.

There is a particular value of a for which the curve has gradient 1 at (0, 1). The letter e is used to stand for this value and the function $f(x) = e^x$ is often called **the exponential function** because of its special importance to the subject.

e = 2.718281828... It is an irrational number and so it cannot be written exactly.

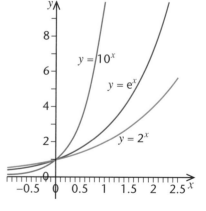

You may be asked to sketch other curves relating to e^x, for example

Notice $e^x > 0$ for all values of x.

Reflection in y-axis.

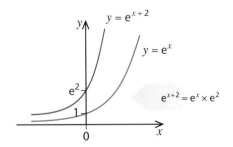

Translation 2 units left or stretch, factor e^2, parallel to y-axis.

The natural logarithmic function, ln x

Core 3

Key points from AS

- **Logarithms**
  Revise AS page 36

Your calculator has a key for natural logs, probably labelled ln.

If $y = e^x$ then $x = \log_e y$. This is usually written as $x = \ln y$.

Logarithms to the base e are called *natural logarithms*.

$e^0 = 1 \Rightarrow \ln 1 = 0$
$e^1 = e \Rightarrow \ln e = 1$

In fact, $\ln e^{f(x)} = f(x)$, for example $\ln e^{x^2} = x^2$, $\ln e^{\sin x} = \sin x$.

This is true for logs of any base.

Since the exponential and natural logarithm functions are inverses of each other, their graphs are symmetrical about the line $y = x$.

The diagram shows that the domain of the natural logarithm function is given by $x > 0$.

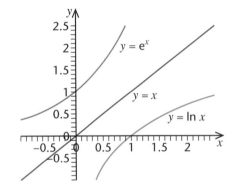

You may be asked to sketch other curves relating to $y = \ln x$, for example

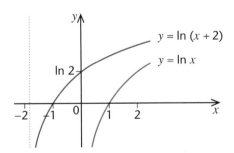

Translation 2 units to the left.

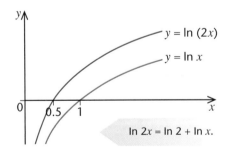

$\ln 2x = \ln 2 + \ln x$.

Stretch parallel to the x-axis factor $\frac{1}{2}$ or translation $\ln 2$ units up.

Progress check

Core 3

1 $f(x) = x^2 + 5$ and $g(x) = 2x + 3$.

   (a) Write $fg(x)$ in terms of x.   (b)   Find $fg(10)$.

   (c) Find the values of x for which $fg(x) = gf(x)$.

Core 3

2 Find the inverse of the function $f(x) = \dfrac{x + 5}{x - 2}$, $x \in \mathbb{R}$, $x \neq 2$ and state the domain of the inverse function.

Core 3

3 (a) Sketch $y = (x - 2)(x + 2)$.

   (b) Sketch $y = |x^2 - 4|$.

Core 3

4 Solve $|x - 2| < 5$.

Core 3

5 The diagram shows $y = f(x)$

On separate diagrams, sketch

   (a) $y = f(x) + 1$

   (b) $y = f(x + 1)$

   (c) $y = f(2x)$

   (d) $y = 2f(x)$

   (e) $y = f(\frac{1}{2}x)$

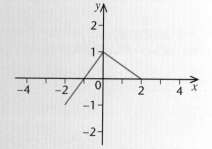

Core 3

6 On the same diagram, sketch $y = x^2$, $y = (x - 2)^2$, $y = -x^2$.

Core 3

7 On the same diagram, sketch $y = e^x$, $y = e^{2x}$, $y = -e^x$.

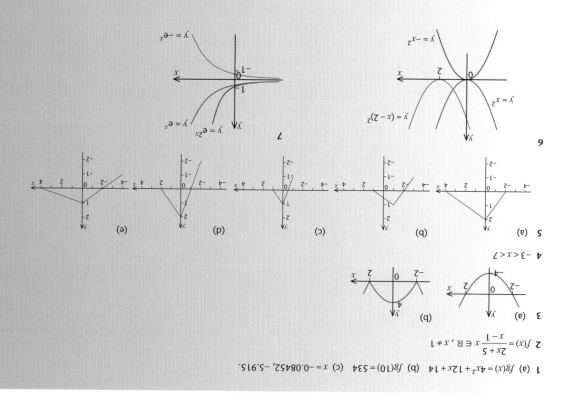

7

6

5 (a) (b) (c) (d) (e)

4 $-3 < x < 7$

3 (a) (b)

2 $f(x) = \dfrac{2x + 5}{x - 1}$, $x \in \mathbb{R}$, $x \neq 1$

1 (a) $fg(x) = 4x^2 + 12x + 14$ (b) $fg(10) = 534$ (c) $x = -0.08452, -5.915$.

1.3 Coordinate geometry

After studying this section you should be able to:

- sketch curves in Cartesian and parametric form
- convert the equation of a curve between Cartesian and parametric form

LEARNING SUMMARY

Cartesian and parametric form

Core 4

Cartesian form

A curve in Cartesian form is defined in terms of x and y only, **for example**
$$y = x^2 + 3x \text{ in which } y \text{ is given explicitly in terms of } x.$$
$$x^2 + 2xy + y^2 = 9 \text{ in which the equation is given implicitly.}$$

Key points from AS

- **Curve sketching**
  Revise AS page 29

If all the powers of x are even, the graph is symmetrical in the y-axis.

Look for values that make the denominator zero.

To sketch a function given in Cartesian form:

1. Find where the curve **crosses the axes** by putting $x = 0$ and $y = 0$.
2. Check for **symmetry**.
3. Check for **stationary points**.
4. Check what happens for **large values** of x and y.
5. Check for **discontinuities**.

Find when $\dfrac{dy}{dx} = 0$ and then investigate.

Parametric form

The parameter θ is often used when trigonometric functions are involved.

The curve is plotted by finding the values of x and y for various values of the parameter.

A curve is defined in **parametric form** by expressing x and y in terms of a third variable, for example $x = f(t)$, $y = g(t)$. In this case, t is the parameter.

Example
Plot the curve $x = 2t^2$, $y = 4t$ for $-3 \leqslant t \leqslant 3$ and find the Cartesian equation of the curve.

The curve is a parabola.

t	−3	−2	−1	0	1	2	3
x	18	8	2	0	2	8	18
y	−12	−8	−4	0	4	8	12

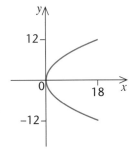

Eliminate t to obtain the Cartesian equation.

$$y = 4t \Rightarrow t = \frac{y}{4}.$$

Substituting into $x = 2t^2$ gives $x = 2\left(\dfrac{y}{4}\right)^2 = \dfrac{y^2}{8}$ so $y^2 = 8x$.

Key points from AS

- **Trigonometric identities**
  Revise AS page 47

When the parameter is θ, it may be necessary to use a trigonometric identity to find the Cartesian equation.

For example, when $x = a \cos \theta$ and $y = a \sin \theta$, use $\cos^2 \theta + \sin^2 \theta = 1$. This gives $\dfrac{x^2}{a^2} + \dfrac{y^2}{a^2} = 1$ or $x^2 + y^2 = a^2$.

This is a circle, centre (0, 0), radius a.

Also, when $x = a \cos \theta$, $y = b \sin \theta$, $\dfrac{x^2}{a^2} + \dfrac{y^2}{b^2} = 1$.

This curve is an ellipse.

Progress check

Core 4

1 Sketch the curve $y = x^3 + 3x^2$.

Core 4

2 A curve is defined parametrically by $x = 2t$, $y = \dfrac{2}{t}$.

   (a) Give the Cartesian equation of the curve.
   (b) Sketch the curve.

Core 4

3 Describe the curve defined parametrically by $x = 5 \cos \theta$, $y = 5 \sin \theta$ and give its Cartesian equation.

3 Circle, centre (0, 0), radius 5; $x^2 + y^2 = 25$.

(b)

2 (a) $xy = 4$ (hyperbola)

1

1.4 Trigonometry

After studying this section you should be able to:

- *understand the functions secant, cosecant and cotangent and be able to draw their graphs*
- *understand and use the Pythagorean identities*
- *understand and use the compound angle formulae for sine, cosine and tangent*
- *use identities to solve equations*

LEARNING SUMMARY

Secant, cosecant and cotangent

Core 3

These trigonometric functions, commonly known as **sec**, **cosec** and **cot**, are defined from the more familiar sin, cos and tan functions as follows:

$$\sec x = \frac{1}{\cos x} \qquad \operatorname{cosec} x = \frac{1}{\sin x} \qquad \cot x = \frac{1}{\tan x}$$

Key points from AS

- **Trigonometry** *pages 44–48*

Each function is periodic and has either line or rotational symmetry. The domain of each one must be restricted to avoid division by zero.
You need to be able to work in degrees or radians.
Remember $180° = \pi$ radians.

$y = \sec x$

- The period of sec is 360° (2π radians) to match the period of cos.
- Notice that sec x is undefined whenever cos $x = 0$.
- The graph is symmetrical about every vertical line passing through a vertex.
- It has rotational symmetry of order 2 about the points in the x-axis corresponding to $90° \pm 180°n$ ($\frac{\pi}{2} \pm \pi n$).

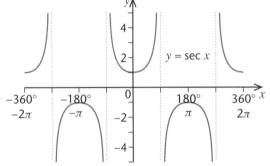

$y = \operatorname{cosec} x$

- The period of cosec is 360° (2π radians) to match the period of sin.
- Notice that cosec x is undefined whenever sin $x = 0$.
- The graph is symmetrical about every vertical line passing through a vertex.
- It has rotational symmetry of order 2 about the points on the x-axis corresponding to $0°$, $\pm 180°$, $\pm 360°$, ..., $(0, \pm \pi, \pm 2\pi, ...)$.

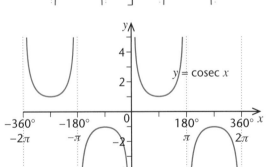

$y = \cot x$

$$\frac{1}{\tan x} = \frac{\cos x}{\sin x}.$$

- The period of cot is 180° to match the period of tan.
- Notice that cot x is undefined whenever sin $x = 0$.
- The graph has rotational symmetry of order 2 about the points on the x-axis corresponding to $0°$, $\pm 90°$, $\pm 180°$, ..., $(0, \pm \frac{\pi}{2}, \pm \pi, ...)$.

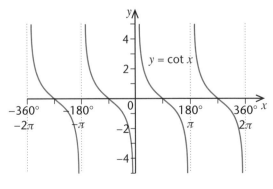

Inverse trigonometric functions

Core 3

The sine, cosine and tangent functions are all many–one and so do not have inverses on their full domains. However, it is possible to restrict their domains so that each one has an inverse. The graphs of these inverse functions are given below.

See page 19 to find out about inverse functions.

You need to know these graphs and to understand the need for restricting the domains.

$f(x) = \sin x, -\frac{\pi}{2} \leqslant x \leqslant \frac{\pi}{2}$
$\quad (-90° \leqslant x \leqslant 90°)$
$\Rightarrow f^{-1}(x) = \sin^{-1}x, -1 \leqslant x \leqslant 1.$

$f(x) = \cos x, 0 \leqslant x \leqslant \pi$
$\quad (0° \leqslant x \leqslant 180°)$
$\Rightarrow f^{-1}(x) = \cos^{-1}x, -1 \leqslant x \leqslant 1.$

$f(x) = \tan x, -\frac{\pi}{2} < x < \frac{\pi}{2}$
$\quad (-90° < x < 90°)$
$\Rightarrow f^{-1}(x) = \tan^{-1}x, x \in \mathbb{R}.$

This is the notation used for inverse trig' functions on your calculator. But some texts use:

arcsin for $\sin^{-1}$

arccos for $\cos^{-1}$

arctan for $\tan^{-1}$.

$\sin^{-1}x$ means:
'the angle whose sine is x'
e.g. $\sin^{-1}0.5 = 30°$.

In radians $\sin^{-1}0.5 = \frac{\pi}{6}$.

$\cos^{-1}x$ means:
'the angle whose cosine is x'
e.g. $\cos^{-1}0.5 = 60°$.

In radians $\cos^{-1}0.5 = \frac{\pi}{3}$.

$\tan^{-1}x$ means:
'the angle whose tangent is x'
e.g. $\tan^{-1}1 = 45°$.

In radians $\tan^{-1}1 = \frac{\pi}{4}$.

When you use the functions $\sin^{-1}$, $\cos^{-1}$ and $\tan^{-1}$ on your calculator, the value given is called the **principal value** (PV).

Trigonometric identities

Core 3

You should *learn* the identities listed below. It is so much easier to simplify expressions, establish new identities and solve equations if you have a firm grasp of these basic results.

The second result can be obtained from the first by dividing throughout by $\cos^2\theta$.
The third result can be obtained in a similar way, by dividing throughout by $\sin^2\theta$.

• The **Pythagorean identities** are:

$$\cos^2\theta + \sin^2\theta \equiv 1$$

$$1 + \tan^2\theta \equiv \sec^2\theta$$

$$\cot^2\theta + 1 \equiv \operatorname{cosec}^2\theta$$

• The **compound angle identities** for $(A + B)$ are:

$$\sin(A + B) \equiv \sin A \cos B + \cos A \sin B$$

$$\cos(A + B) \equiv \cos A \cos B - \sin A \sin B$$

$$\tan(A + B) \equiv \frac{\tan A + \tan B}{1 - \tan A \tan B}.$$

The corresponding results for $(A - B)$ are:

$$\sin(A - B) \equiv \sin A \cos B - \cos A \sin B$$

$$\cos(A - B) \equiv \cos A \cos B + \sin A \sin B$$

These results can be worked out from the ones for $(A + B)$. Just replace every '+' with a '–' and vice versa. This makes it easier to learn the results.

$$\tan(A - B) \equiv \frac{\tan A - \tan B}{1 + \tan A \tan B}.$$

You can obtain these results from the compound angle formulae by setting $A = B$.

- The **double angle identities** are:

$$\sin 2A \equiv 2 \sin A \cos A$$

$$\cos 2A \equiv \cos^2 A - \sin^2 A$$

$$\equiv 2 \cos^2 A - 1$$

$$\equiv 1 - 2 \sin^2 A$$

These variations are very useful – it is worth knowing them.

$$\tan 2A \equiv \frac{2 \tan A}{1 - \tan^2 A}.$$

- The **half angle identities** are:

These results will be useful for integration later in the course.

$$\sin^2 \tfrac{1}{2} A \equiv \tfrac{1}{2}(1 - \cos A)$$

$$\cos^2 \tfrac{1}{2} A \equiv \tfrac{1}{2}(1 + \cos A).$$

Proving identities

Core 3

The basic technique for proving a given identity is:

- Start with the expression on one side of the identity.
- Use known identities to replace some part of it with an equivalent form.
- Simplify the result and compare with the expression on the other side.

This is a skill. The more you do, the better you will become.

With practice you get to know which parts to replace so that the required result is established.

Example

Prove that $\dfrac{\cos 2A}{\cos A - \sin A} \equiv \cos A + \sin A.$

Starting with the LHS:

$$\frac{\cos 2A}{\cos A - \sin A} \equiv \frac{\cos^2 A - \sin^2 A}{\cos A - \sin A}$$

Using one of the double angle results.

$$\equiv \frac{(\cos A + \sin A)(\cos A - \sin A)}{\cos A - \sin A}$$

Using the difference of two squares.

$$\equiv \cos A + \sin A \equiv \text{RHS}.$$

Cancelling the common factor.

It is often a good idea to replace sec, cosec and cot with sin, cos and tan at the start.

Example

Prove that $\sec^2 A \equiv \dfrac{\text{cosec } A}{\text{cosec } A - \sin A}.$

It's often easier to start with the more complicated looking side and try to simplify it.

$$\text{RHS} \equiv \frac{\dfrac{1}{\sin A}}{\dfrac{1}{\sin A} - \sin A} \equiv \frac{1}{1 - \sin^2 A}$$

Replacing cosec A with $\dfrac{1}{\sin A}$ and multiplying the numerator and denominator by sin A.

$$\equiv \frac{1}{\cos^2 A} \equiv \sec^2 A \equiv \text{LHS}.$$

Using one of the Pythagorean identities and the definition of sec A.

Trigonometric equations

Core 3

Identities may be used to re-write an equation in a form that is easier to solve.

Example Solve the equation $\cos 2x + \sin x + 1 = 0$ for $-\pi \leqslant x \leqslant \pi$.

Replacing $\cos 2x$ with $1 - 2 \sin^2 x$ gives $1 - 2 \sin^2 x + \sin x + 1 = 0$

$$\Rightarrow 2 \sin^2 x - \sin x - 2 = 0.$$

> You need to recognise that this is a quadratic in $\sin x$.

Using the quadratic formula gives $\sin x = \dfrac{1 \pm \sqrt{1 + 16}}{4} = \dfrac{1 \pm \sqrt{17}}{4}$

$$\Rightarrow \sin x = 1.28 \ldots \text{ (no solutions) or } \sin x = -0.78077 \ldots .$$

Using $\sin^{-1}$ gives $\qquad x = -0.8959$ radians to 4 d.p.

> Set calculator to radian mode.

> $-1 \leqslant \sin x \leqslant 1$ for all values of x.

Another solution in the interval $-\pi \leqslant x \leqslant \pi$ is $-\pi + 0.8959$ giving $x = -2.2457$ radians to 4 d.p.

The diagram shows there are no other solutions in this interval.

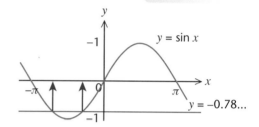

> **KEY POINT**
>
> An important result is that the expression $a \cos \theta + b \sin \theta$ can be written in any of the equivalent forms $r \cos(\theta \pm \alpha)$ or $r \sin(\theta \pm \alpha)$ for a suitable choice of r and α. The convention is to take $r > 0$.

One application of the result is to solve equations of the form $a \cos \theta + b \sin \theta = c$.

Example Express $3 \cos \theta - 4 \sin \theta$ in the form $r \cos(\theta + \alpha)$, $r > 0$. Use the result to solve the equation $3 \cos \theta - 4 \sin \theta = 2$ for $-\pi \leqslant \theta \leqslant \pi$.

> You need to expand $r \cos(\theta + \alpha)$ and match corresponding parts of the identity.

$$r \cos(\theta + \alpha) \equiv 3 \cos \theta - 4 \sin \theta$$

$$\Rightarrow r \cos \theta \cos \alpha - r \sin \theta \sin \alpha \equiv 3 \cos \theta - 4 \sin \theta$$

$$\Rightarrow r \cos \alpha = 3 \text{ and } r \sin \alpha = 4$$

> $\dfrac{r \sin \alpha}{r \cos \alpha} = \dfrac{4}{3}$
>
> $\Rightarrow \tan \alpha = \dfrac{4}{3}$

$$\Rightarrow r = \sqrt{3^2 + 4^2} = 5 \text{ and } \alpha = \tan^{-1}\left(\tfrac{4}{3}\right) = 0.9273^c \text{ to 4 d.p.}$$

$$\Rightarrow 3 \cos \theta - 4 \sin \theta \equiv 5 \cos(\theta + 0.9273^c).$$

Using this result in the equation $3 \cos \theta - 4 \sin \theta = 2$ gives

$$5 \cos(\theta + 0.9273^c) = 2$$

$$\Rightarrow \cos(\theta + 0.9273^c) = 0.4$$

This gives $\theta + 0.9273^c = \pm 1.1593^c \Rightarrow \theta = 0.232^c$ or $\theta = -2.087^c$ to 3 d.p.

Progress check

Core 3

1 Prove the identity $\sin 2\theta \equiv \dfrac{2 \tan \theta}{1 + \tan^2 \theta}$.

Core 3

2 Solve the equation $\cos 2x - \sin x = 0$ for $0° \leqslant x \leqslant 360°$.

Core 3

3 (a) Write $5 \sin x + 12 \cos x$ in the form $r \sin(x + \alpha)$ where $0 < \alpha < \dfrac{\pi}{2}$.

   (b) Solve the equation $5 \sin x + 12 \cos x = 7$ for $-\pi \leqslant x \leqslant \pi$.

3 (a) $13 \sin(x + 1.176^c)$ (b) $x = -0.607^c$ or $x = 1.397^c$ to 3 d.p.

2 $x = 30°, 150°$ or $270°$

1 RHS $= \dfrac{2 \tan \theta}{\sec^2 \theta} = 2 \tan \theta \times \cos^2 \theta = 2 \sin \theta \cos \theta \equiv \sin 2\theta \equiv$ LHS.

1.5 Differentiation

After studying this section you should be able to:

- use the chain rule to differentiate composite functions
- differentiate functions based on e^x and $\ln x$
- differentiate trigonometric functions
- differentiate using the product and quotient rules
- understand and use the relation $\dfrac{dy}{dx} = 1 \div \dfrac{dx}{dy}$
- find derivatives of functions defined parametrically and implicitly

LEARNING SUMMARY

The chain rule

Core 3

> **KEY POINT**
>
> The **chain rule** can be written as $\dfrac{dy}{dx} = \dfrac{dy}{du} \times \dfrac{du}{dx}$, where u is a function of x.

It shows how to differentiate a composite function by separating the process into simpler stages and combining the results.

Example   Differentiate   (a) $y = (3x+2)^{10}$

(b) $y = \sqrt{x^2 - 9}$.

Key points from AS

- **Differentiation**
  Revise AS pages 28–31

(a) Taking $u = 3x + 2$ gives $y = u^{10}$ and $\dfrac{dy}{du} = 10u^9$, so $\dfrac{dy}{du} = 10(3x+2)^9$.

It also gives $\dfrac{du}{dx} = 3$.

So, from the chain rule: $\dfrac{dy}{dx} = 10(3x+2)^9 \times 3 = 30(3x+2)^9$.

(b) Taking $u = x^2 - 9$ gives $y = \sqrt{u} = u^{\frac{1}{2}}$, and $\dfrac{dy}{du} = \frac{1}{2}u^{-\frac{1}{2}}$, so $\dfrac{dy}{du} = \frac{1}{2}(x^2-9)^{-\frac{1}{2}}$.

It also gives $\dfrac{du}{dx} = 2x$.

So, from the chain rule: $\dfrac{dy}{dx} = \frac{1}{2}(x^2-9)^{-\frac{1}{2}} \times 2x = \dfrac{x}{\sqrt{x^2-9}}$.

The chain rule can also be used to establish results for **connected rates of change**.

For example, the rate of change of the volume of a sphere can be written as $\dfrac{dv}{dt}$.

The corresponding rate of change of the radius of the sphere can be written as $\dfrac{dr}{dt}$. Using the chain rule, the connection between these rates of change is given by:

$$\frac{dv}{dt} = \frac{dv}{dr} \times \frac{dr}{dt}.$$

Also, since $v = \frac{4}{3}\pi r^3$, it follows that $\dfrac{dv}{dr} = 4\pi r^2$, so, the connection can now be written as $\dfrac{dv}{dt} = 4\pi r^2 \dfrac{dr}{dt}$.

The exponential function e^x

Core 3

The exponential function was defined on page 22. It has the special property that it is unchanged by differentiation. So, if $y = e^x$ then $\dfrac{dy}{dx} = e^x$.

Constant multiples of the exponential function behave in the same way, so if

$y = ae^x$ then $\dfrac{dy}{dx} = ae^x$ where a is any constant.

Examples

Differentiate with respect to x:

(a) $y = 2e^x$ (b) $y = x^3 + e^x$ (c) $f(x) = \dfrac{1}{x} - 3e^x$

(a) $\dfrac{dy}{dx} = 2e^x$ (b) $\dfrac{dy}{dx} = 3x^2 + e^x$ (c) $f'(x) = -\dfrac{1}{x^2} - 3e^x$

You may need to use the chain rule.

Examples

Differentiate with respect to x:

(a) $y = e^{3x+2}$ (b) $y = e^{4x^2}$

(a) Let $u = 3x + 2$ so that $y = e^u$

$$\dfrac{du}{dx} = 3 \quad \text{and} \quad \dfrac{dy}{du} = e^u$$

Using the chain rule

$$\dfrac{dy}{dx} = \dfrac{dy}{du} \times \dfrac{du}{dx} = e^u \times 3 = 3\,e^{3x+2}$$

(b) Let $u = 4x^2$, so that $y = e^u$

$$\dfrac{du}{dx} = 8x \quad \text{and} \quad \dfrac{dy}{du} = e^u$$

$$\dfrac{dy}{dx} = \dfrac{dy}{du} \times \dfrac{du}{dx} = e^u \times 8x = 8xe^{4x^2}$$

In general, it can be shown using the chain rule

> $f'(x)$ is the derivative, with respect to x, of $f(x)$.

KEY POINT

If $y = e^{f(x)}$ then $\dfrac{dy}{dx} = f'(x)e^{f(x)}$.

In particular, it is useful to remember that if $y = e^{ax+b}$, then $\dfrac{dy}{dx} = ae^{ax+b}$

Logarithmic functions $\ln x$ and $\ln f(x)$

Core 3

KEY POINT

If $y = \ln x$ then $\dfrac{dy}{dx} = \dfrac{1}{x}$.

Key points from AS

• **Logarithms**
  Revise AS page 36

$\log a + \log b = \log(ab).$
$\log x^k = k \log x.$

This result is easily extended to functions of the form $y = \ln(kx)$ and $y = \ln(x^k)$ using the laws of logarithms.

If $y = \ln(kx)$, then $y = \ln k + \ln x$

giving $\dfrac{dy}{dx} = \dfrac{1}{x} + 0 = \dfrac{1}{x}.$

> Since $\ln k$ is a constant, its derivative is zero.

For example, if $y = \ln(5x)$, then $\dfrac{dy}{dx} = \dfrac{1}{x}.$

If $y = \ln(x^k)$, then $y = k \ln x$ giving $\dfrac{dy}{dx} = k \times \dfrac{1}{x} = \dfrac{k}{x}.$

For example, if $y = \ln\left(\dfrac{1}{x^2}\right) = \ln(x^{-2}) = -2 \ln x$, then $\dfrac{dy}{dx} = -2 \times \dfrac{1}{x} = -\dfrac{2}{x}.$

Always check whether a logarithmic function can be simplified before attempting to differentiate.

You may need to use the chain rule to differentiate $y = \ln f(x).$

Example

> This function cannot be simplified.

Differentiate $y = \ln(x^2 + 3x).$

Let $u = x^2 + 3x$ so that $y = \ln u$

$\dfrac{du}{dx} = 2x + 3$ and $\dfrac{dy}{du} = \dfrac{1}{u} = \dfrac{1}{x^2 + 3x}$

Using the chain rule

$\dfrac{dy}{dx} = \dfrac{dy}{du} \times \dfrac{du}{dx} = \dfrac{1}{x^2 + 3x} \times (2x + 3)$

> Notice that $2x + 3$ is the derivative of $x^2 + 3x.$

$= \dfrac{2x + 3}{x^2 + 3x}$

In general, it can be shown using the chain rule

> This can be used directly without showing the chain rule working.

$$\frac{d}{dx}\ln(f(x)) = \frac{f'(x)}{f(x)}.$$

KEY POINT

Progress check

Find $\dfrac{dy}{dx}$ in each of the following questions.

Core 3

1 $y = (4 - 3x)^6$

Core 3

2 $y = \sqrt{2x + 1}$

Core 3

3 $y = e^{3x+1} + 3x$

Core 3

4 $y = \ln(7x) + \ln(x^4)$

Core 3

5 $y = \ln(x^2 - 3)$

5 $\dfrac{2x}{x^2 - 3}$

4 $\dfrac{5}{x}$

3 $3e^{3x+1} + 3$

2 $(2x + 1)^{-\frac{1}{2}}$

1 $-18(4 - 3x)^5$

Trigonometric functions

Core 4

You should learn the derivatives of the three main **trigonometric functions**:

$$\frac{d}{dx}(\sin x) = \cos x \qquad \frac{d}{dx}\sin(ax+b) = a\cos(ax+b)$$

$$\frac{d}{dx}(\cos x) = -\sin x \qquad \frac{d}{dx}\cos(ax+b) = -a\sin(ax+b)$$

$$\frac{d}{dx}(\tan x) = \sec^2 x \qquad \frac{d}{dx}\tan(ax+b) = a\sec^2(ax+b)$$

> When differentiating, the angle must be in radians.

Example

Find the exact value of the gradient of $y = \cos 2x$ when $x = \dfrac{\pi}{6}$.

$$y = \cos 2x \implies \frac{dy}{dx} = -2\sin 2x$$

$$\text{When } x = \frac{\pi}{6}, \quad \frac{dy}{dx} = -2\sin\frac{\pi}{3} - -2 \times \frac{\sqrt{3}}{2} = -\sqrt{3}$$

The **chain rule** can often be used to differentiate expressions involving trigonometric functions.

Example

Differentiate: (a) $y = 5\sin^4 x$ (b) $y = e^{\tan 4x}$ (c) $y = \ln(\sin x)$

> $\sin^4 x = (\sin x)^4$.

(a)   Let $u = \sin x$ so that $y = 5u^4$

Then $\dfrac{du}{dx} = \cos x$ and $\dfrac{dy}{du} = 20u^3 = 20(\sin x)^3 = 20\sin^3 x$

> With practice you should be able to do the working mentally.

By the chain rule:

$$\frac{dy}{dx} = \frac{dy}{du} \times \frac{du}{dx} = 20\sin^3 x \times (\cos x)$$

$$= 20\cos x \sin^3 x.$$

> Although not essential, it is usual to write an expression with the least complicated factors first.

(b)   Let $u = \tan 4x$ so that $y = e^u$

$$\frac{du}{dx} = 4\sec^2 4x \text{ and } \frac{dy}{du} = e^u = e^{\tan 4x}.$$

$$\therefore \quad \frac{dy}{dx} = e^{\tan 4x} \times 4\sec^2 4x$$

$$= 4\sec^2 4x\, e^{\tan 4x}.$$

> In general, by the chain rule:
> $$\frac{d}{dx}(e^{f(x)}) = f'(x)e^{f(x)}.$$

(c)   Use the result $y = \ln f(x) \implies \dfrac{dy}{dx} = \dfrac{f'(x)}{f(x)}$

$$y = \ln(\sin x)$$

$$\frac{dy}{dx} = \frac{\cos x}{\sin x}$$

$$= \cot x.$$

> $f(x) = \sin x,\ f'(x) = \cos x.$

The product rule

Core 3

Expressions written as the product of two functions, $f(x) \times g(x)$, can be differentiated using the **product rule**.

> **KEY POINT**
>
> If $y = uv$, where $u = f(x)$ and $v = g(x)$, then $\dfrac{dy}{dx} = u\dfrac{dv}{dx} + v\dfrac{du}{dx}$.

Example

Differentiate with respect to x: (a) $y = x^2 \sin 4x$ (b) $y = x^3 e^{-x}$

Example (a)
Core 4

(a) Let $y = uv$, where $u = x^2$ and $v = \sin 4x$ Side working

$$\frac{dy}{dx} = u\frac{dv}{dx} + v\frac{du}{dx}$$

$$u = x^2 \Rightarrow \frac{du}{dx} = 2x$$

With practice, the side working can be done mentally.

$$= x^2 \times 4\cos 4x + \sin 4x \times 2x$$

$$v = \sin 4x \Rightarrow \frac{dv}{dx} = 4\cos 4x$$

$$= 2x(2x\cos 4x + \sin 4x)$$

Tidy the answer by taking out factors if possible.

(b) Let $y = uv$, where $u = x^3$ and $v = e^{-x}$ Side working

$$\frac{dy}{dx} = x^3 \times (-e^{-x}) + e^{-x} \times 3x^2$$

$$u = x^3 \Rightarrow \frac{du}{dx} = 3x^2$$

$$= x^2 e^{-x}(-x + 3)$$

$$v = e^{-x} \Rightarrow \frac{dv}{dx} = -e^{-x}$$

The quotient rule

Core 3

Expressions written in the form $\dfrac{f(x)}{g(x)}$ can be differentiated using the **quotient rule**.

> **KEY POINT**
>
> If $y = \dfrac{u}{v}$ where $u = f(x)$ and $v = g(x)$, then $\dfrac{dy}{dx} = \dfrac{v\dfrac{du}{dx} - u\dfrac{dv}{dx}}{v^2}$.

Before using the quotient rule, check whether the expression can be simplified first, such as in

$$y = \frac{4x^2 + 3x + 2}{x}.$$

Example

Find $\dfrac{dy}{dx}$ when $y = \dfrac{e^{2x}}{x^3}$.

Side working:

$$\frac{dy}{dx} = \frac{v\dfrac{du}{dx} - u\dfrac{dv}{dx}}{v^2}$$

$$u = e^{2x} \Rightarrow \frac{du}{dx} = 2e^{2x}$$

$$= \frac{x^3(2e^{2x}) - e^{2x}(3x^2)}{(x^3)^2}$$

$$v = x^3 \Rightarrow \frac{dv}{dx} = 3x^2$$

Take out factors.

$$= \frac{x^2 e^{2x}(2x - 3)}{x^6}$$

Simplify.

$$= \frac{e^{2x}(2x - 3)}{x^4}$$

Example

Find the coordinates of the stationary point on the curve $y = \dfrac{\ln x}{x}$.

Side working:

$$\frac{dy}{dx} = \frac{x \times \dfrac{1}{x} - \ln x \times 1}{x^2}$$

$$= \frac{1 - \ln x}{x^2}$$

$u = \ln x \Rightarrow \dfrac{du}{dx} = \dfrac{1}{x}$

$v = x \Rightarrow \dfrac{dv}{dx} = 1$

$\dfrac{dy}{dx} = 0$ when $\ln x = 1$, i.e. when $x = e$

In e = 1.

When $x = e$, $y = e^{-1}$, so the stationary point is at (e, e^{-1}).

Alternative formats of product and quotient rules when $u = f(x)$ and $v = g(x)$:

These formats are sometimes quoted. Learn and use the ones you find easiest to remember.

Product rule: If $y = f(x) \times g(x)$, then $\dfrac{dy}{dx} = f'(x)g(x) + f(x)g'(x)$

Quotient rule: If $y = \dfrac{f(x)}{g(x)}$, then $\dfrac{dy}{dx} = \dfrac{f'(x)g(x) - f(x)g'(x)}{[g(x)]^2}$

Using the result $\dfrac{dy}{dx} = 1 \div \dfrac{dx}{dy}$

Core 3

$$\frac{dy}{dx} = 1 \div \frac{dx}{dy} = \frac{1}{\dfrac{dx}{dy}}.$$

KEY POINT

For example, if $x = 4y + 2$, then $\dfrac{dx}{dy} = 4$ and $\dfrac{dy}{dx} = \dfrac{1}{\dfrac{dx}{dy}} = \dfrac{1}{4}$.

Parametric functions

Core 4

If x and y are each expressed in terms of a **parameter** t,

then $\dfrac{dy}{dx} = \dfrac{dy}{dt} \times \dfrac{dt}{dx}$.   Remember that $\dfrac{dt}{dx} = \dfrac{1}{dx/dt}$.

KEY POINT

Example

A curve is defined parametrically by $x = 3t^2$, $y = 4t$.

Find $\dfrac{dy}{dx}$ when $y = 8$

$$x = 3t^2 \Rightarrow \frac{dx}{dt} = 6t \Rightarrow \frac{dt}{dx} = \frac{1}{6t}$$

$$y = 4t \Rightarrow \frac{dy}{dt} = 4$$

$$\therefore \quad \frac{dy}{dx} = \frac{dy}{dt} \times \frac{dt}{dx} = 4 \times \frac{1}{6t} = \frac{2}{3t}$$

Find the value of t when $y = 8$.

When $y = 8$, $4t = 8$, i.e. $t = 2$; when $t = 2$, $\dfrac{dy}{dx} = \dfrac{2}{3t} = \dfrac{2}{6} = \dfrac{1}{3}$.

Implicit functions

Core 4

> To find $\dfrac{dy}{dx}$ when an equation in x and y is given **implicitly**, differentiate each term with respect to x, remembering that $\dfrac{d}{dx}(f(y)) = \dfrac{d}{dy}f(y) \times \dfrac{dy}{dx}$.

Example

The point $P(2, 6)$ lies on the circle $x^2 + y^2 + 2x - 4y - 20 = 0$.

(a) Show that $\dfrac{dy}{dx} = \dfrac{x + 1}{2 - y}$

(b) Calculate the gradient of the tangent at P.

(a) $2x + 2y\dfrac{dy}{dx} + 2 - 4\dfrac{dy}{dx} + 0 = 0$

Differentiate term by term with respect to x

$\dfrac{d}{dx}(y) = \dfrac{dy}{dx}$.

$x + y\dfrac{dy}{dx} + 1 - 2\dfrac{dy}{dx} = 0$

Divide throughout by 2 to simplify the equation.

Get the terms in $\dfrac{dy}{dx}$ on one side of the equation and all the other terms on the other side.

$x + 1 = 2\dfrac{dy}{dx} - y\dfrac{dy}{dx}$

Take out $\dfrac{dy}{dx}$ as a factor.

$\Rightarrow x + 1 = \dfrac{dy}{dx}(2 - y)$

$\therefore \quad \dfrac{dy}{dx} = \dfrac{x + 1}{2 - y}$

(b) At P, $x = 2$ and $y = 6$,

so $\dfrac{dy}{dx} = \dfrac{3}{-4} = -\dfrac{3}{4}$.

The value of $\dfrac{dy}{dx}$ when $x = 2$ gives the gradient of the tangent at $(2, 6)$

The gradient of the tangent at P is $-\dfrac{3}{4}$.

Progress check

Core 4

1 Differentiate with respect to x:

   (a) $y = 2 \sin 6x$ (b) $y = 4 \cos^3 x$ (c) $y = 2 \tan 3x$

   (d) $y = e^{\sin x}$ (e) $y = \sin(x^3)$

Core 4

2 If $y = \sin^2 x$, show that $\dfrac{dy}{dx} = \sin 2x$.

> Use a trigonometric identity.

Core 3

Part (c)
Core 4

3 Differentiate with respect to x:

   (a) $\ln(x^2 + 3)$ (b) $\ln(2x - 4)^5$ (c) $\ln(\cos 2x)$ (d) $\ln(6x - 1)$

4 Find $\dfrac{dy}{dx}$, simplifying your answers.

Core 4

   (a) $y = x^2 \cos 3x$

   (b) $y = e^{2x} \sin x$

Core 3

   (c) $y = x \ln x$

   (d) $y = (x + 1)(2x - 3)^4$

Core 3

5 Find the equation of the tangent to the curve $y = x\sqrt{1 + 2x}$ at the point (4, 12).

Core 3

6 Use the quotient rule to find $\dfrac{dy}{dx}$, simplifying your answers where necessary.

Part (a)
Core 4

   (a) $y = \dfrac{\sin 2x}{x}$ (b) $y = \dfrac{2x - 1}{3x + 4}$ (c) $y = \dfrac{\ln(x + 2)}{x + 2}$

Core 4

7 A curve is defined parametrically by $x = t^2$, $y = t^3$.

   (a)   Find $\dfrac{dy}{dx}$ when $t = 2$.

   (b)   Find the Cartesian equation of the curve.

Core 4

8 A curve has equation $x^2 + xy^2 - 5 = 0$.

   (a)   Find the value of $\dfrac{dy}{dx}$ at (1, 2).

   (b)   Find the equation of the normal when $x = 1$, in the form $ax + by + c = 0$.

8 (a) -1.5 (b) $2x - 3y + 4 = 0$

7 (a) 3 (b) $x^3 = y^2$

6 (a) $\dfrac{2x \cos 2x - \sin 2x}{x^2}$ (b) $\dfrac{11}{(3x + 4)^2}$ (c) $\dfrac{1 - \ln(x + 2)}{(x + 2)^2}$

5 $3y = 13x - 16$

4 (a) $x(2 \cos 3x - 3x \sin 3x)$ (b) $e^{2x}(\cos x + 2 \sin x)$ (c) $1 + \ln x$ (d) $5(2x + 1)(2x - 3)^3$

3 (a) $\dfrac{2x}{x^2 + 3}$ (b) $\dfrac{5}{x - 2}$ (c) $-2 \tan 2x$ (d) $\dfrac{6x}{6x - 1}$

1 (a) $12 \cos 6x$ (b) $-12 \sin x \cos^2 x$ (c) $6 \sec^2 3x$ (d) $\cos x \, e^{\sin x}$ (e) $3x^2 \cos(x^3)$

1.6 Integration

After studying this section you should be able to:

- *integrate* e^x *and simple variations of this function with respect to* x
- *integrate trigonometrical functions*
- *recognise an integral that leads to a logarithmic function*
- *integrate using partial fractions*
- *integrate using a substitution*
- *use integration by parts to integrate a product*
- *use integration to find a volume of revolution*
- *solve first order differential equations when the variables can be separated*
- *understand exponential growth and decay*

LEARNING SUMMARY

The exponential function e^x

Core 3

The fact that differentiation leaves the exponential function unchanged was stated on page 32. Since integration is the reverse process of differentiation it follows that

> **KEY POINT**
> $$\int e^x \, dx = e^x + c \quad \text{and} \quad \int a e^x \, dx = a \int e^x \, dx = a e^x + c.$$

An extension of this result, using the reverse of the chain rule, is

> **KEY POINT**
> $$\int e^{ax+b} \, dx = \frac{1}{a} e^{ax+b} + c.$$

Examples

1. $\int 4e^x \, dx = 4e^x + c$

2. $\int \frac{1}{2} e^{3x-1} \, dx = \frac{1}{6} e^{3x-1} + c$

Trigonometric functions

Core 4

The following integrals are obtained by applying the reverse process of differentiating the three main trigonometric functions (page 34).

You should learn these and practise using them.

$$\int \cos x \, dx = \sin x + c \qquad \int \cos(ax+b) \, dx = \frac{1}{a} \sin(ax+b) + c$$

$$\int \sin x \, dx = -\cos x + c \qquad \int \sin(ax+b) \, dx = -\frac{1}{a} \cos(ax+b) + c$$

$$\int \sec^2 x \, dx = \tan x + c \qquad \int \sec^2(ax+b) \, dx = \frac{1}{a} \tan(ax+b) + c$$

Example

It is a good idea to check your answer by differentiating.

$$\int (4 \cos 2x - \sin 2x) \, dx = 2 \sin 2x + \tfrac{1}{2} \cos 2x + c$$

Example

When integrating, the angle must be in radians.

Learn the exact values of the trigonometric ratios of $\frac{\pi}{6}, \frac{\pi}{4}, \frac{\pi}{3}$ (30°, 45°, 60°).

$$\int_{\frac{\pi}{3}}^{\frac{\pi}{2}} \sec^2(\tfrac{1}{2}x)dx = \left[\frac{1}{\frac{1}{2}} \tan(\tfrac{1}{2}x)\right]_{\frac{\pi}{3}}^{\frac{\pi}{2}}$$

$$= 2\left[\tan\left(\tfrac{1}{2}x\right)\right]_{\frac{\pi}{3}}^{\frac{\pi}{2}}$$

$$= 2\left(\tan\frac{\pi}{4} - \tan\frac{\pi}{6}\right)$$

$$= 2\left(1 - \frac{1}{\sqrt{3}}\right)$$

These results are very useful and can be recognised easily:

$$\frac{d}{dx}(\sin^{n+1} x) = (n+1)\sin^n x \cos x \implies \int \sin^n x \cos x\, dx = \frac{1}{n+1}\sin^{n+1} x + c$$

$$\frac{d}{dx}(\cos^{n+1} x) = -(n+1)\cos^n x \sin x \implies \int \cos^n x \sin x\, dx = -\frac{1}{n+1}\cos^{n+1} x + c$$

Examples

$$\int \sin^5 x \cos x\, dx = \tfrac{1}{6}\sin^6 x + c$$

$$\int \sin x \cos^3 x\, dx = -\tfrac{1}{4}\cos^4 x + c$$

Odd powers of sin x and cos x

Example

Evaluate $\displaystyle\int_0^{\frac{\pi}{6}} \cos^3 x\, dx$

Use $\cos^2 x = 1 - \sin^2 x$.

Recognise the integral of $\cos x \sin^2 x$.

$$\int_0^{\frac{\pi}{6}} \cos^3 x\, dx = \int_0^{\frac{\pi}{6}} \cos x(\cos^2 x)dx$$

$$= \int_0^{\frac{\pi}{6}} \cos x(1 - \sin^2 x)dx$$

$$= \int_0^{\frac{\pi}{6}} (\cos x - \cos x \sin^2 x)dx$$

$$= \left[\sin x - \tfrac{1}{3}\sin^3 x\right]_0^{\frac{\pi}{6}}$$

$$= \tfrac{1}{2} - \tfrac{1}{24} = \tfrac{11}{24}$$

Even powers of cos x and sin x

To integrate even powers of $\sin x$ and $\cos x$, use the double angle identities for $\cos 2x$.

$$\cos 2x = 2\cos^2 x - 1 \implies \cos^2 x = \tfrac{1}{2}(1 + \cos 2x)$$

$$\cos 2x = 1 - 2\sin^2 x \implies \sin^2 x = \tfrac{1}{2}(1 - \cos 2x)$$

Example

$$\int \cos^2 x \, dx = \frac{1}{2} \int (1 + \cos 2x) dx$$

$$= \frac{1}{2}(x + \frac{1}{2} \sin 2x) + c$$

To integrate $\cos^4 x$, write it as $(\cos^2 x)^2 = (\frac{1}{2}(1 + \cos 2x))^2$.

The reciprocal function $\frac{1}{x}$

Core 3

The result $\int x^n \, dx = \dfrac{x^{n+1}}{n+1} + c$ holds for any value of n other than $n = -1$.

In this special case, the result takes a different form:

Remember that $x^{-1} = \dfrac{1}{x}$.

> **KEY POINT**
>
> $$\int \frac{1}{x} \, dx = \ln|x| + c.$$

This extends to $\displaystyle\int \frac{k}{x} \, dx = k \int \frac{1}{x} \, dx = k \ln|x| + c,$

and to $\displaystyle\int \frac{1}{kx} \, dx = \frac{1}{k} \int \frac{1}{x} \, dx = \frac{1}{k} \ln|x| + c.$ \qquad (where k is a constant)

For example, $\displaystyle\int \frac{2}{x} \, dx = 2 \ln|x| + c$ and $\displaystyle\int \frac{1}{3x} \, dx = \frac{1}{3} \ln|x| + c.$

Integrals leading to a logarithmic function

Core 4

Remember that $\dfrac{d}{dx} \ln(f(x)) = \dfrac{f'(x)}{f(x)}.$

Applying the process in reverse gives:

> **KEY POINT**
>
> $$\int \frac{f'(x)}{f(x)} \, dx = \ln |f(x)| + c.$$

Examples

(a) $\displaystyle\int \frac{x^2}{2x^3 - 2} \, dx = \frac{1}{6} \ln | 2x^3 - 2 | + c$

$\dfrac{d}{dx}(2x^3 - 2) = 6x^2$ so a factor of $\frac{1}{6}$ is needed.

(b) $\displaystyle\int \tan x \, dx = \int \frac{\sin x}{\cos x} \, dx$

$$= -\ln | \cos x | + c$$

$$= \ln | \sec x | + c$$

$-\ln a = \ln(a^{-1}).$

Care must be taken with **definite integrals** of this type.
These can be evaluated between the limits a and b only if $f(x)$ exists for all values between a and b. Note that $f(a)$ and $f(b)$ will have the same sign.

$$\int_a^b \frac{f'(x)}{f(x)} \, dx = \left[\ln|f(x)| \right]_a^b = (\ln|f(b)| - \ln|f(a)|)$$

Example

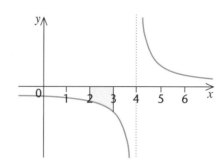

The diagram shows the curve $y = \dfrac{1}{x-4}$.

Find the area between the curve, the x-axis, and the lines $x = 2$ and $x = 3$.

$$\int_a^b y \, \mathrm{d}x = \int_2^3 \frac{1}{x-4} \, \mathrm{d}x$$

$$= \left[\ln|x-4| \right]_2^3$$

$\ln|-2| = \ln 2$
$\ln 1 = 0$.

$$= \ln|-1| - \ln|-2|$$

$$= \ln 1 - \ln 2$$

$$= -\ln 2 \ (= -0.69 \ldots)$$

Note that the formula gives a negative value, confirming that the area is below the x-axis.

Shaded area $= \ln 2 \ (= 0.69 \ldots)$ square units.

$f(3)$ and $f(5)$ do not have the same sign.

Note that $\displaystyle\int_3^5 \frac{1}{x-4} \, \mathrm{d}x$ cannot be evaluated, as the curve is undefined at $x = 4$.

Partial fractions

Core 4

Some rational functions can be integrated by expressing them in partial fraction form.

Example

Find $\displaystyle\int \frac{x+7}{(x-2)(x+1)} \, \mathrm{d}x$, simplifying your answer.

Decompose into partial fractions (see page 15).

$$\frac{x+7}{(x-2)(x+1)} \equiv \frac{3}{x-2} - \frac{2}{x+1}$$

The working for this example is shown on page 15.

$$\Rightarrow \int \frac{x+7}{(x-2)(x+1)} \, \mathrm{d}x = \int \left(\frac{3}{x-2} - \frac{2}{x+1} \right) \mathrm{d}x$$

$$= 3\ln(x-2) - 2\ln(x+1) + c$$

$$= \ln k \frac{(x-2)^3}{(x+1)^2}$$

$\log a - \log b = \log\left(\dfrac{a}{b}\right)$

$\log a^n = n \log a$.

Note: Sometimes the constant c is incorporated into the log function by letting $c = \ln k$.

Substitution

Core 4

Sometimes an integral is made easier by using a substitution.

This transforms the integral with respect to one variable, say x, into an integral with respect to a related variable, say u.

$$\int f(x)\,\mathrm{d}x = \int f(x)\,\frac{\mathrm{d}x}{\mathrm{d}u}\,\mathrm{d}u.$$

This method is known as the method of **substitution** or **change of variable**.

> This could also be done by using the substitution $u = 2x + 1$.

Example

Find $\int x\sqrt{2x + 1}\,\mathrm{d}x$, using the substitution $u = \sqrt{2x + 1}$.

$$\int x\sqrt{2x + 1}\,\mathrm{d}x = \int x\sqrt{2x + 1}\,\frac{\mathrm{d}x}{\mathrm{d}u}\,\mathrm{d}u \qquad\qquad \text{Side working}$$

> Show the side working as part of your answer.

$$= \int \tfrac{1}{2}(u^2 - 1)u \times u\,\mathrm{d}u \qquad\qquad u = \sqrt{2x + 1}$$

$$= \frac{1}{2}\int (u^4 - u^2)\,\mathrm{d}u \qquad\qquad u^2 = 2x + 1 \implies x = \tfrac{1}{2}(u^2 - 1)$$

> When simplifying, take out factors as soon as possible.

$$= \frac{1}{2}\left(\frac{u^5}{5} - \frac{u^3}{3}\right) + c \qquad\qquad \frac{\mathrm{d}x}{\mathrm{d}u} = u$$

$$= \frac{u^3}{2}\left(\frac{u^2}{5} - \frac{1}{3}\right) + c$$

> Write the answer in terms of x, simplifying as much as possible.

$$= \tfrac{1}{30}u^3(3u^2 - 5) + c \qquad\qquad 3u^2 - 5 = 3(2x + 1) - 5$$

$$= \tfrac{1}{30}(2x + 1)^{\frac{3}{2}}(6x - 2) + c \qquad\qquad = 6x - 2$$

$$= \tfrac{1}{15}(2x + 1)^{\frac{3}{2}}(3x - 1) + c \qquad\qquad = 2(3x - 1)$$

When evaluating definite integrals using the method of substitution, change the x limits to u limits and substitute them into the working as soon as possible. There is no need to tidy up the expression first.

Sometimes it is possible to bypass the method of substitution by **recognising** an integral as the derivative of a particular function.

Examples

> This type of integral can always be done by using a substitution.

(a) $\int (2x + 1)^5\,\mathrm{d}x = \tfrac{1}{12}(2x + 1)^6 + c$

> Recognise that
> $$\frac{\mathrm{d}}{\mathrm{d}x}(2x + 1)^6 = 12(2x + 1)^5.$$

(b) $\int x\sqrt{2x^2 + 1}\,\mathrm{d}x = \tfrac{1}{6}(2x^2 + 1)^{\frac{3}{2}} + c$

> Recognise that
> $$\frac{\mathrm{d}}{\mathrm{d}x}(2x^2 + 1)^{\frac{3}{2}} = 6x(2x^2 + 1)^{\frac{1}{2}}.$$

General result:

$$\int f'(x)[f(x))]^n\,\mathrm{d}x = \frac{1}{n + 1}[f(x)]^{n+1} + c$$

Integration by parts

Core 4

It is sometimes possible to integrate a product using **integration by parts**.

> KEY POINT
>
> If u and v are functions of x, then $\displaystyle\int u\,\frac{\mathrm{d}v}{\mathrm{d}x}\,\mathrm{d}x = uv - \int v\,\frac{\mathrm{d}u}{\mathrm{d}x}\,\mathrm{d}x.$

Before using integration by parts, check whether the function can be simplified, or whether the integral can be recognised directly or done using a substitution.

Example

Find $\displaystyle\int 2x \sin x\,\mathrm{d}x.$

Let $u = 2x$, then $\dfrac{\mathrm{d}u}{\mathrm{d}x} = 2$

Let $\dfrac{\mathrm{d}v}{\mathrm{d}x} = \sin x$ then $v = -\cos x$

Integration by parts can only be used if v can be found from $\dfrac{\mathrm{d}v}{\mathrm{d}x}$ and it is possible to find $\displaystyle\int v\,\frac{\mathrm{d}u}{\mathrm{d}x}\,\mathrm{d}x.$

$$\therefore \int 2x \sin x\,\mathrm{d}x = 2x(-\cos x) - \int (-\cos x \times 2)\,\mathrm{d}x$$

$$= -2x \cos x + \int 2 \cos x\,\mathrm{d}x$$

$$= -2x \cos x + 2 \sin x + c$$

Sometimes the integration by parts process has to be carried out more than once. This happens when integrating expressions such as $x^2 e^x$ or $x^2 \cos x$.

Definite integrals

$$\int_a^b u\,\frac{\mathrm{d}v}{\mathrm{d}x}\,\mathrm{d}x = \left[uv\right]_a^b - \int_a^b v\,\frac{\mathrm{d}u}{\mathrm{d}x}\,\mathrm{d}x$$

Example

$$\int_0^1 xe^{2x}\,\mathrm{d}x = \left[x\tfrac{1}{2}e^{2x}\right]_0^1 - \int_0^1 \tfrac{1}{2}e^{2x}\,\mathrm{d}x$$

$$= \tfrac{1}{2}e^2 - \left[\tfrac{1}{4}e^{2x}\right]_0^1$$

$$= \tfrac{1}{2}e^2 - (\tfrac{1}{4}e^2 - \tfrac{1}{4}) = \tfrac{1}{4}e^2 + \tfrac{1}{4}$$

Working:

Let $u = x$, then $\dfrac{\mathrm{d}u}{\mathrm{d}x} = 1$

Let $\dfrac{\mathrm{d}v}{\mathrm{d}x} = e^{2x}$, then $v = \tfrac{1}{2}e^{2x}$

Some functions containing $\ln x$ can be integrated using integration by parts. A useful strategy is to take $\ln x$ as u.

Often a polynomial function is taken as u because it becomes of lower degree when differentiated. Integrals involving $\ln x$ are an exception, since $\ln x$ is not easy to integrate, but can be differentiated.

Example

$$\int x^2 \ln x\,\mathrm{d}x = \int (\ln x \times x^2)\,\mathrm{d}x$$

$$= \ln x \times \left(\frac{x^3}{3}\right) - \int \left(\frac{x^3}{3} \times \frac{1}{x}\right)\mathrm{d}x$$

$$= \tfrac{1}{3}x^3 \ln x - \tfrac{1}{3}\int x^2\,\mathrm{d}x$$

$$= \tfrac{1}{3}x^3 \ln x - \tfrac{1}{9}x^3 + c$$

Working:

Let $u = \ln x$, then $\dfrac{\mathrm{d}u}{\mathrm{d}x} = \dfrac{1}{x}$

Let $\dfrac{\mathrm{d}v}{\mathrm{d}x} = x^2$, then $v = \dfrac{x^3}{3}$

This method is particularly useful when integrating $\ln x$.
Think of it as integrating the product $\ln x \times 1$ as follows:

$$\int (\ln x \times 1)dx = \ln x \times x - \int \left(x \times \frac{1}{x}\right)dx$$

Working:

Let $u = \ln x$, then $\dfrac{du}{dx} = \dfrac{1}{x}$

$$= x \ln x - \int 1 dx$$

Let $\dfrac{dv}{dx} = 1$, then $v = x$

$$= x \ln x - x + c$$

Volume of revolution

Core 3

The shaded region R is bounded by the curve $y = f(x)$, the x-axis and the lines $x = a$ and $x = b$.

Rotating R completely about the x-axis forms a solid figure. The volume of this figure is called the volume of revolution and is given by

$$V = \int_a^b \pi y^2 \ dx$$

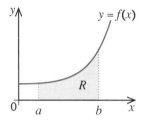

The corresponding result for rotation about the y-axis is

$$\int_a^b \pi x^2 \ dy$$

where the region is bounded by the curve $y = f(x)$, the y-axis and the lines $y = a$ and $y = b$.

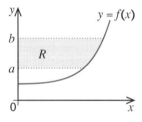

Example Find a formula for the volume of a cone of height h and base radius r.

The volume of the cone is given by the volume of revolution of the shaded region shown.

The straight line has gradient $\dfrac{r}{h}$ and passes through the origin, so its equation is $y = \dfrac{r}{h}x$.

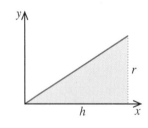

Using the formula for volume of revolution gives:

$$V = \int_0^h \pi \left(\frac{r}{h}x\right)^2 dx = \frac{\pi r^2}{h^2} \int_0^h x^2 \ dx$$

$$= \frac{\pi r^2}{h^2} \left[\frac{x^3}{3}\right]_0^h$$

$$= \frac{\pi r^2}{h^2} \times \frac{h^3}{3}$$

$$= \frac{\pi r^2 h}{3}.$$

Differential equations

A **first order differential equation** in x and y contains only the first differential coefficient.

$$\text{Examples} \quad \frac{dy}{dx} = -2x, \qquad \frac{dy}{dx} = \frac{x}{y}, \qquad y\frac{dy}{dx} = 3$$

If $\dfrac{dy}{dx} = -2x$, then integrating with respect to x gives $y = -x^2 + c$.

> The general solution contains an arbitrary integration constant.

This is the **general solution** of the differential equation and it can be illustrated by a family of curves.

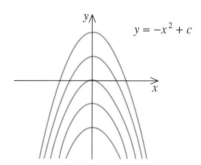

$$y = -x^2 + c$$

A **particular solution** is a specific member of the family and it can be found from additional information.

> The value of c is found to give a particular solution.

For example, if the curve $y = -x^2 + c$ goes through the point $(2, 0)$, then $0 = -(2)^2 + c \Rightarrow c = 4$. The particular solution is $y = -x^2 + 4$.

Separating the variables

> **KEY POINT**
>
> A differential equation that can be written in the form $f(y)\dfrac{dy}{dx} = g(x)$
>
> can be solved by **separating the variables**, where $\displaystyle\int f(y)dy = \int g(x)dx$.

Example

Solve (a) $\dfrac{dy}{dx} = \dfrac{x}{y}$ \qquad (b) $\dfrac{dy}{dx} = y$

(a) $\qquad \dfrac{dy}{dx} = \dfrac{x}{y}$

> Separate the variables to get the format $\int \ldots dy = \int \ldots dx$.

$$\int y\, dy = \int x\, dx \Rightarrow \tfrac{1}{2}y^2 = \tfrac{1}{2}x^2 + c$$

(b) $\qquad \dfrac{dy}{dx} = y$

> This is a special case, when $g(x) = 1$.

$$\frac{1}{y}\frac{dy}{dx} = 1$$

$$\int \frac{1}{y}\, dy = \int 1\, dx$$

> The answer can be left like this.

$$\ln y = x + c$$
$$y = e^{x+c} = e^x \times e^c = Ae^x \quad (\text{where } A = e^c)$$

$$\text{So } \frac{dy}{dx} = y \Rightarrow y = Ae^x$$

> The letter A is often used here, but it could be any letter (other than e, x or y, of course!).

Exponential growth and decay

Core 4

You may have to form a differential equation from given information. Often this models exponential growth or exponential decay.

Exponential growth	Exponential decay
	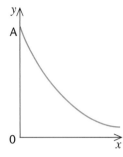

In exponential decay, the fact that y is decreasing with respect to x is usually shown by the negative, keeping $k > 0$.

$$\frac{dy}{dx} = ky \Rightarrow y = Ae^{kx} \quad (k > 0)$$

$$\frac{dy}{dx} = -ky \Rightarrow y = Ae^{-kx} \quad (k > 0)$$

p_0 is the initial population.

An example of exponential growth is the growth of a population where

$$\frac{dp}{dt} = kt \Rightarrow p = p_0 e^{-kt}$$

Examples of exponential decay are:

m_0 is the initial mass.

(a) the disintegration of radioactive materials, $\dfrac{dm}{dt} = -km \Rightarrow m = m_0 e^{-kt}$

T_s is temperature of surroundings, T_0 is initial temperature of the object.

(b) Newton's Law of Cooling, $\dfrac{dT}{dt} = -k(T - T_s) \Rightarrow T - T_s = (T_0 - T_s)e^{-kt}$

Core 3 and Core 4 (Pure Mathematics)

Progress check

Core 3

1 Integrate with respect to x.

(a) $\dfrac{3}{x}$ (b) $\dfrac{x+3}{x^2}$ (c) $3e^x - \dfrac{1}{x}$ (d) $3e^{4-x}$

Core 3

2 The region bounded by the curve $y = e^{\frac{x}{2}}$, the lines $x = 1$ and $x = 3$ and the x-axis is rotated completely about the x-axis. Find the volume of revolution.

Core 4

3 (a) $\displaystyle\int 4\sin 2x\,dx$ (b) $\displaystyle\int \sec^2 2x\,dx$ (c) $\displaystyle\int \cos\left(\tfrac{1}{2}x\right)dx$

(d) $\displaystyle\int 4\sin^2 x\,dx$ (e) $\displaystyle\int \cos^2 2x\,dx$ (f) Evaluate $\displaystyle\int_0^{\frac{1}{6}\pi} \sin^4 x\cos x\,dx$

Part (a) Core 3

4 Find (a) $\displaystyle\int \dfrac{1}{2x+5}\,dx$

Core 4

(b) $\displaystyle\int \cot x\,dx$

(c) $\displaystyle\int \dfrac{3x+2}{x^2-x-12}\,dx$

> Split (c) into partial fractions first.

Core 4

5 Using the substitution $u = 2 + e^{2x}$, or otherwise, find $\displaystyle\int \dfrac{e^{2x}}{2+e^{2x}}\,dx$

> Remember that $\dfrac{dx}{du} = \dfrac{1}{du/dx}$.

Core 4

6 (a) Find (i) $\displaystyle\int xe^{-x}\,dx$ (ii) $\displaystyle\int x^3 \ln x\,dx$ (b) Evaluate $\displaystyle\int_0^{\frac{1}{2}\pi} x\cos x\,dx$

Core 4

7 Solve the differential equation $\dfrac{dy}{dx} = 2xy$, given that $y = 3$ when $x = 0$.

---

7 $y = 3e^{x^2}$

6 (a) (i) $-e^{-x}(x+1) + c$ (ii) $\frac{1}{16}x^4(4\ln x - 1) + c$ (b) $\frac{1}{2}\pi - 1$

5 $\frac{1}{2}\ln(2 + e^{2x}) + c$

4 (a) $\frac{1}{2}\ln(2x+5) + c$ (b) $\ln(\sin x) + c$ (c) $\ln(x+3)(x-4)^2 + c$ or $\ln A(x+3)(x-4)^2$

3 (a) $-2\cos 2x + c$ (b) $\frac{1}{2}\tan 2x + c$ (c) $2\sin\left(\frac{1}{2}x\right) + c$

(d) $2x - \sin 2x + c$ (e) $\frac{1}{2}x + \frac{1}{8}\sin 4x + c$ (f) $\frac{1}{160}$

2 $\displaystyle\int_1^3 \pi e^x\,dx = \pi[e^x]_1^3 = \pi(e^3 - e)$.

1 (a) $3\ln|x| + c$ (b) $\ln|x| - \frac{3}{x} + c$ (c) $3e^x - \ln|x| + c$ (d) $-3e^{4-x} + c$

48

1.7 Numerical methods

After studying this section you should be able to:

- locate an interval containing a root of an equation by finding a change of sign
- use simple iteration to solve equations
- find the approximate value of an integral using Simpson's rule

LEARNING SUMMARY

Change of sign

Core 3

If there is an interval from $x = a$ to $x = b$ in which the graph of a function has no breaks then the function is said to be **continuous** on the interval.

All polynomial functions are continuous everywhere.

For example, the graph of $y = x^2 - 3$ is continuous everywhere.

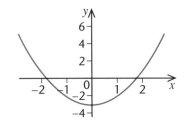

However, the graph of $y = \dfrac{1}{x - 3}$ is not

continuous on any interval that contains the value $x = 3$.

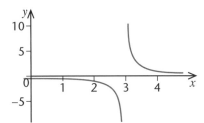

> **KEY POINT**
>
> If $f(x)$ is continuous between $x = a$ and $x = b$ and if $f(a)$ and $f(b)$ have different signs, then a root of $f(x) = 0$ lies in the interval from a to b.

Example Given that $f(x) = e^x - 10x$, show that the equation $f(x) = 0$ has a root between 3 and 4.

$f(3) = -9.914\ldots < 0$

$f(4) = 14.598\ldots > 0$.

The change of sign shows that $f(x) = 0$ has a root between 3 and 4.

You can continue this process, called a **decimal search**, to trap the root in a smaller and smaller interval. The table gives the values of the function in steps of 0.1.

Most graphic calculators can tabulate values of a function and this gives a very efficient way to obtain the information.

x	3.1	3.2	3.3	3.4	3.5	3.6	3.7	3.8	3.9
$f(x)$	−8.802	−7.467	−5.887	−4.035	−1.884	0.5982	3.4473	6.7011	10.402

This shows that the root lies between 3.5 and 3.6
At the mid-point $f(3.55) = -0.6866\ldots < 0$ and so the root lies between 3.55 and 3.6

This is a good way to establish the value of a root to a particular level of accuracy.

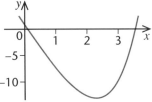

It follows that the value of the root is 3.6 to 1 d.p.

Iteration

There are different methods for producing an iterative formula. You only need to know about simple iteration for this component.

To solve an equation by **iteration** you start with some approximation to a root and improve its accuracy by substituting it into a formula. You can then repeat the process until you have the desired level of accuracy.

Using **simple iteration**, the first step is to rearrange the equation to express x as a function of itself. This function defines the iterative formula that you need.

For example, the equation $e^x - 10x = 0$ can be rearranged as

$$e^x = 10x$$

$$\Rightarrow x = \ln(10x).$$

This may now be turned into an iterative formula for finding x.

$$x_{n+1} = \ln(10x_n).$$

The values may be found very easily using a calculator's Ans function:
Key in 3 =
followed by
ln(10Ans) = = =
to produce the sequence.
(Depending on the make of your calculator you may need to use EXE or ENTER in place of = .)

Starting with $x_1 = 3$, $x_2 = \ln 30 = 3.401\ldots$ and so on. This produces a sequence of values:

$x_2 = 3.401\ldots$ $x_6 = 3.5760\ldots$

$x_3 = 3.526\ldots$ $x_7 = 3.5768\ldots$

$x_4 = 3.563\ldots$ $x_8 = 3.5770\ldots$

$x_5 = 3.573\ldots$ $x_9 = 3.5771\ldots$

The calculator display soon settles on 3.577152064. The value of the root is $x = 3.577$ to 4 s.f.

The diagram shows how the process converges from the starting point on either side of the root.

Start from a point on the x-axis, move up to the curve and across to $y = x$. Then move up to the curve and across to $y = x$ again. Continue in the same way. Each of these stages corresponds to one iteration of the formula.

A different rearrangement of the original equation gives $x = \dfrac{e^x}{10}$ and so the corresponding iterative formula is $x_{n+1} = \dfrac{e^{x_n}}{10}$.

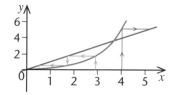

You may need to try several rearrangements in order to converge to a particular root of an equation by this method.

Starting with $x = 3$, this rearrangement fails to converge to the root between 3 and 4 but it does converge to the other root of the equation.

Notice how the movement is away from the upper root this time. If the starting value is smaller than the upper root then the process converges to the lower root. A starting value above the upper root fails to converge to either root.

The value of this root is $x = 0.1118$ to 4 s.f.

Example Starting with $x_1 = 1$, use the formula $x_{n+1} = \dfrac{x_n^2 - 5x_n}{10} - 3$ to find $x_2, x_3, \ldots, x_8$ and describe the long term behaviour of the sequence.

$x_2 = -3.4$

$x_3 = -0.144$

$x_4 = -2.9259264$

$x_5 = -0.6809322$

$x_6 = -2.6131669$

$x_7 = -1.0105523$

$x_8 = -2.3926032$

The sequence oscillates but converges to −1.787 to 4 s.f. which is the lower root of the equation

$$x = \frac{x^3 - 5x}{10} - 3,$$

which simplifies to

$$x^2 - 15x - 30 = 0.$$

The oscillations correspond to a cobweb pattern on the diagram.

Numerical integration

Core 3

Simpson's rule

This method divides the area into an **even** number of parallel strips and approximates the areas of pairs of strips using the following formula:

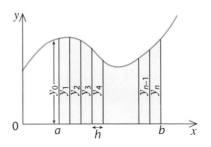

$$\int_a^b f(x)dx \approx \tfrac{1}{3}h\{(y_0 + y_n) + 4(y_1 + y_3 + \ldots + y_{n-1}) + 2(y_2 + y_4 + \ldots + y_{n-2})\}$$

where $h = \dfrac{b-a}{n}$ and n is even.

Note that the first ordinate has been labelled y_0 and the last y_n. There are n strips and $n+1$ ordinates.

Example

Estimate $\displaystyle\int_0^1 e^{x^2}dx$ using Simpson's rule with 10 strips.

Tabulating the results helps in doing the final calculation.

x	y	First and last ordinates	Odd ordinates	Other ordinates
0	y_0	1		
0.1	y_1		1.010 ...	
0.2	y_2			1.040 ...
0.3	y_3		1.094 ...	
0.4	y_4			1.173 ...
0.5	y_5		1.284 ...	
0.6	y_6			1.433 ...
0.7	y_7		1.632 ...	
0.8	y_8			1.896 ...
0.9	y_9		2.247 ...	
1	y_{10}	2.718 ...		
Totals		3.718 ...	7.268 ...	5.544 ...

If you have multiple memories in your calculator, retain the figures. Otherwise approximate say to 4 decimal places.

$h = \dfrac{1-0}{10} = 0.1$

$$\int_a^b f(x)dx \approx \tfrac{1}{3}h\{(y_0 + y_{10}) + 4(y_1 + y_3 + y_5 + y_7 + y_9) + 2(y_2 + y_4 + y_6 + y_8)\}$$

$$= \tfrac{1}{3} \times 0.1 \times \{3.718 \ldots + 4(7.268 \ldots) + 2(5.544 \ldots)\}$$

$$= 1.463 \; (3 \text{ s.f.})$$

Progress check

Core 3

1 Show that the equation $x^3 - x - 7 = 0$ has a root between 2 and 3. Use a decimal search to find the value of the root to 2 d.p.

Core 3

2 Use simple iteration to refine the value of the root found in question 1 and state its value to 4 d.p.

Core 3

3 Estimate $\int_0^1 \sqrt{1-x^3}\, \mathrm{d}x$ using Simpson's rule with 4 strips.

3 To 3 d.p. 0.823

2 2.0867

1 2.09

1.8 Vectors

After studying this section you should be able to:

- *understand the distinction between vector and scalar quantities*
- *add and subtract vector quantities and multiply by a scalar*
- *use the unit vectors **i**, **j** and **k***
- *find the magnitude and direction of a vector*
- *use scalar products to find the angle between two directions*
- *express equations of lines in vector form*
- *determine whether two lines are parallel, intersect or are skew*

Vector and scalar quantities

Core 4

A **scalar** quantity has size (or **magnitude**) but not direction. **Numbers** are scalars and some other important examples are **distance**, **speed**, **mass** and **time**.

A **vector** quantity has both size and **direction**. For example, distance in a specified direction is called **displacement**.

The diagram shows a **directed line segment**. It has size (in this case, length) and direction so it is a vector.

The diagram gives a useful way to represent *any* vector quantity and may also be used to represent addition and subtraction of vectors and multiplication of a vector by a scalar.

In a textbook, vectors are usually labelled with lower case letters in bold print.
When hand-written, these letters should be underlined e.g. *a*.

Addition and subtraction of vectors

Core 4

This diagram shows three vectors **a**, **b** and **c** such that **c** = **a** + **b**.

The vectors **a** and **b** follow on from each other and then **c** joins the start of vector **a** to the end of vector **b**.

Check that **a** + **b** = **b** + **a**.

The vector **c** is called the **resultant** of **a** and **b**.

On a vector diagram, −**q** has the opposite sense of direction to **q**.
Notice that **p** − **q** is represented as **p** + (−**q**).

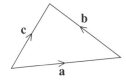

This diagram shows the same information in a different way.

Following the route in the opposite direction to **q** is the same as adding −**q** or subtracting **q**.

Scalar multiplication

Core 4

2**p** is parallel to **p** and has the same sense of direction but is twice as long.

−3**p** is parallel to **p** but has the opposite sense of direction and is three times as long.

Component form

Core 4

When working in two dimensions, it is often very useful to express a vector in terms of two special vectors **i** and **j**. These are **unit vectors** at right-angles to each other. A vector **r** written as $\mathbf{r} = a\mathbf{i} + b\mathbf{j}$ is said to have **components** $a\mathbf{i}$ and $b\mathbf{j}$.

> A unit vector is a vector of magnitude 1 unit.

Examples

$\mathbf{r} = \mathbf{i} + 2\mathbf{j}$

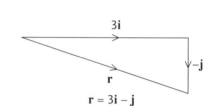

$\mathbf{r} = 3\mathbf{i} - \mathbf{j}$

For work involving the Cartesian coordinate system, **i** and **j** are taken to be in the positive directions of the x- and y-axes respectively.

In three dimensions, a third vector **k** is used to represent a unit vector in the positive direction of the z-axis.

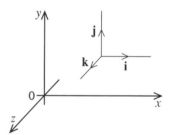

The **position vector** of a point P is the vector $\overrightarrow{OP}$ where O is the origin.

If P has coordinates (a, b) then its position vector is given by $\mathbf{r} = a\mathbf{i} + b\mathbf{j}$.

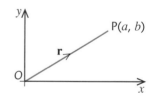

In three dimensions, a point with coordinates (a, b, c) would have position vector $\mathbf{r} = a\mathbf{i} + b\mathbf{j} + c\mathbf{k}$.

Adding and subtracting vectors in component form

Core 4

One advantage of expressing vectors in terms of **i**, **j** and **k** is that addition and subtraction may be done algebraically.

Example
$\mathbf{p} = 3\mathbf{i} + 2\mathbf{j} - \mathbf{k}$ and $\mathbf{q} = 5\mathbf{i} - \mathbf{j} + 4\mathbf{k}$. Express the following vectors in terms of **i**, **j** and **k**:

(a) $\mathbf{p} + \mathbf{q}$ (b) $\mathbf{p} - \mathbf{q}$ (c) $2\mathbf{p} - 3\mathbf{q}$.

(a) $\mathbf{p} + \mathbf{q} = (3\mathbf{i} + 2\mathbf{j} - \mathbf{k}) + (5\mathbf{i} - \mathbf{j} + 4\mathbf{k}) = 8\mathbf{i} + \mathbf{j} + 3\mathbf{k}$

(b) $\mathbf{p} - \mathbf{q} = (3\mathbf{i} + 2\mathbf{j} - \mathbf{k}) - (5\mathbf{i} - \mathbf{j} + 4\mathbf{k}) = -2\mathbf{i} + 3\mathbf{j} - 5\mathbf{k}$

(c) $2\mathbf{p} - 3\mathbf{q} = 2(3\mathbf{i} + 2\mathbf{j} - \mathbf{k}) - 3(5\mathbf{i} - \mathbf{j} + 4\mathbf{k})$
$= 6\mathbf{i} + 4\mathbf{j} - 2\mathbf{k} - 15\mathbf{i} + 3\mathbf{j} - 12\mathbf{k} = -9\mathbf{i} + 7\mathbf{j} - 14\mathbf{k}$

Finding the magnitude of a vector

Core 4

If $\mathbf{r} = a\mathbf{i} + b\mathbf{j} + c\mathbf{k}$, the magnitude (length) of $\mathbf{r}$ is given by

$$|\mathbf{r}| = \sqrt{a^2 + b^2 + c^2}$$

For example,
If $\mathbf{r} = 3\mathbf{i} - 4\mathbf{j} + 12\mathbf{k}$, then $|\mathbf{r}| = \sqrt{3^2 + (-4)^2 + 12^2} = 13$.

Scalar product

Core 4

If $\mathbf{a} = x_1\mathbf{i} + y_1\mathbf{j} + z_1\mathbf{k}$, then $|\mathbf{a}| = \sqrt{x_1^2 + y_1^2 + z_1^2}$ where $|\mathbf{a}|$ is the length of $\mathbf{a}$.

If $\mathbf{b} = x_2\mathbf{i} + y_2\mathbf{j} + z_2\mathbf{k}$, then $|\mathbf{b}| = \sqrt{x_2^2 + y_2^2 + z_2^2}$.

If θ is the angle between $\mathbf{a}$ and $\mathbf{b}$,
the **scalar** (or **dot**) **product** of $\mathbf{a}$ and $\mathbf{b}$ is $\mathbf{a}.\mathbf{b}$
where $\mathbf{a}.\mathbf{b} = |\mathbf{a}||\mathbf{b}|\cos\theta$.

This leads to the useful results:
$\mathbf{i}.\mathbf{i} = 1$, $\mathbf{j}.\mathbf{j} = 1$, $\mathbf{k}.\mathbf{k} = 1$
$\mathbf{i}.\mathbf{j} = \mathbf{j}.\mathbf{k} = \mathbf{k}.\mathbf{i} = 0$.

If two vectors are **parallel**, $\theta = 0$ and $\mathbf{a}.\mathbf{b} = |\mathbf{a}||\mathbf{b}|$.
If two vectors are **perpendicular**, $\theta = 90°$ and $\mathbf{a}.\mathbf{b} = 0$.

For $\mathbf{a}$ and $\mathbf{b}$ defined as above:

This uses the results for parallel and perpendicular vectors.

In column vectors

$$\mathbf{a}.\mathbf{b} = \begin{pmatrix} x_1 \\ y_1 \\ z_1 \end{pmatrix} \cdot \begin{pmatrix} x_2 \\ y_2 \\ z_2 \end{pmatrix}$$
$$= x_1 x_2 + y_1 y_2 + z_1 z_2.$$

$\mathbf{a}.\mathbf{b} = (x_1\mathbf{i} + y_1\mathbf{j} + z_1\mathbf{k}).(x_2\mathbf{i} + y_2\mathbf{j} + z_2\mathbf{k})$
$\qquad = x_1 x_2 + y_1 y_2 + z_1 z_2$

Example
Find the angle between $\mathbf{a} = 2\mathbf{i} + \mathbf{j} + 4\mathbf{k}$ and $\mathbf{b} = -3\mathbf{i} + 2\mathbf{j} - \mathbf{k}$.

$\mathbf{a}.\mathbf{b} = 2(-3) + 1(2) + 4(-1) = -8$
$|\mathbf{a}| = \sqrt{2^2 + 1^2 + 4^2} = \sqrt{21}$
$|\mathbf{b}| = \sqrt{(-3)^2 + 2^2 + (-1)^2} = \sqrt{14}$
$\mathbf{a}.\mathbf{b} = |\mathbf{a}||\mathbf{b}|\cos\theta$
$-8 = \sqrt{21}\sqrt{14}\cos\theta$
$\Rightarrow \cos\theta = \dfrac{-8}{\sqrt{21}\sqrt{14}} = -0.4665\ldots$
$\Rightarrow \theta = 117.8°$ (1 d.p.)

In column vectors:

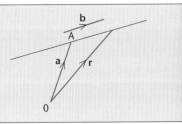

Vector equation of a line

Core 4

A **vector equation** of a line passing
through a fixed point A with position
vector $\mathbf{a}$ and parallel to a vector $\mathbf{b}$ is
$\mathbf{r} = \mathbf{a} + t\mathbf{b}$, where t is a scalar
parameter.

t is often used for the parameter, but not exclusively so. Other letters sometimes used include s, μ and λ.

r is the position vector of any point on the line.
b is the **direction vector** of the line.

The equation of the line is not unique. Any other point on the line could be used instead of A and any multiple of **b** could be used for the direction vector.

In column format:
$$\mathbf{r} = \begin{pmatrix} 4 \\ -5 \\ 1 \end{pmatrix} + t \begin{pmatrix} 3 \\ 4 \\ -2 \end{pmatrix}.$$

Example

A vector equation of the line passing through $(4, -5, 1)$, parallel to $3\mathbf{i} + 4\mathbf{j} - 2\mathbf{k}$ is $\mathbf{r} = 4\mathbf{i} - 5\mathbf{j} + \mathbf{k} + t(3\mathbf{i} + 4\mathbf{j} - 2\mathbf{k})$.

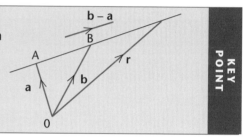

KEY POINT

A vector equation of the line through two fixed points A and B, with position vectors **a** and **b** is given by $\mathbf{r} = \mathbf{a} + t(\mathbf{b} - \mathbf{a})$ where t is a scalar parameter.

Example

Find a vector equation of the line through the points $P(3, 1, -4)$ and $Q(-2, 5, 1)$.

Direction of line PQ:

$$\overrightarrow{PQ} = \mathbf{q} - \mathbf{p} = -2\mathbf{i} + 5\mathbf{j} + \mathbf{k} - (3\mathbf{i} + \mathbf{j} - 4\mathbf{k}) = -5\mathbf{i} + 4\mathbf{j} + 5\mathbf{k}$$

A vector equation of PQ is $\mathbf{r} = 3\mathbf{i} + \mathbf{j} - 4\mathbf{k} + t(-5\mathbf{i} + 4\mathbf{j} + 5\mathbf{k})$.

Angles between two lines

To find the angle between two lines, find the angle between their direction vectors.

Pairs of lines

Core 4

In 2-dimensions, a pair of lines are either parallel or they intersect.
In 3-dimensions, a pair of lines are parallel, or they intersect or they are **skew**.

If two lines are parallel, then their direction vectors are multiples of each other, as in this **example**:
$\mathbf{r} = 3\mathbf{i} - 5\mathbf{j} + 2\mathbf{k} + \lambda(\mathbf{i} + 2\mathbf{j} - \mathbf{k})$ and $\mathbf{r} = 4\mathbf{i} + \mathbf{j} - 2\mathbf{k} + \mu(3\mathbf{i} + 6\mathbf{j} - 3\mathbf{k})$.

The two lines $\mathbf{r} = \mathbf{a} + \lambda\mathbf{b}$ and $\mathbf{r} = \mathbf{c} + \mu\mathbf{d}$ intersect if unique values of λ and μ can be found such that $\mathbf{a} + \lambda\mathbf{b} = \mathbf{c} + \mu\mathbf{d}$.
If unique values cannot be found, then the two lines are **skew**.

Example

Investigate whether the lines $\mathbf{r}_1 = 2\mathbf{i} + \mathbf{j} - 3\mathbf{k} + \lambda(4\mathbf{i} + 6\mathbf{j} - \mathbf{k})$ and $\mathbf{r}_2 = 3\mathbf{i} - 2\mathbf{k} + \mu(4\mathbf{i} + \mathbf{j} + 3\mathbf{k})$ intersect or whether they are skew.

If the lines intersect, then $\mathbf{r}_1 = \mathbf{r}_2$ will have a unique solution for λ and μ.

Note that the lines are not parallel, since their direction vectors are not multiples of each other.

$\mathbf{r}_1 = \mathbf{r}_2 \implies 2\mathbf{i} + \mathbf{j} - 3\mathbf{k} + \lambda(4\mathbf{i} + 6\mathbf{j} - \mathbf{k}) = 3\mathbf{i} - 2\mathbf{k} + \mu(4\mathbf{i} + \mathbf{j} + 3\mathbf{k})$

i.e. $(2 + 4\lambda)\mathbf{i} + (1 + 6\lambda)\mathbf{j} + (-3 - \lambda)\mathbf{k} = (3 + 4\mu)\mathbf{i} + (0 + \mu)\mathbf{j} + (-2 + 3\mu)\mathbf{k}$

Equating coefficients:

Solving (1) and (2) simultaneously gives:

$2 + 4\lambda = 3 + 4\mu$	(1)	$\lambda = -0.25$ and $\mu = -0.5$
$1 + 6\lambda = \mu$	(2)	Check whether these satisfy (3):
$-3 - \lambda = -2 + 3\mu$	(3)	LHS $= -3 - (-0.25) = -2.75$
		RHS $= -2 + 3(-0.5) = -3.5$

∴ $\lambda = -0.25$ and $\mu = -0.5$ do not satisfy all three equations.

Since there is not a unique solution for λ and μ, the lines do not intersect. They are skew.

Progress check

Core 4

1 Write down the position vector **r** of a point with coordinates (5, −2).

Core 4

2 Find the magnitude of the vector **r** = 5**i** − 12**j** and give its direction relative to **i**.

Core 4

3 Find the angle between the vectors **a** = 2**i** + 3**j** − **k** and **b** = **i** + **j** + 2**k**.

Core 4

4 (a) Give a vector equation of the line which is parallel to 4**i** − **j** + 2**k** and goes through the point with coordinates (3, 0, 1).

(b) Give a vector equation of the line through (4, −1, 1) and (−3, 2, −2).

Core 4

5 (a) Find the point of intersection of the lines

$$\mathbf{r} = 2\mathbf{i} - 3\mathbf{j} + \mu(\mathbf{i} - 2\mathbf{j}) \text{ and } \mathbf{r} = \mathbf{i} - 2\mathbf{j} + \lambda(2\mathbf{i} + 3\mathbf{j}).$$

(b) Investigate whether the following lines intersect or whether they are skew. If they intersect, find their point of intersection.

$$\mathbf{r} = \mathbf{i} + 2\mathbf{j} - 5\mathbf{k} + s(3\mathbf{i} + \mathbf{j} + \mathbf{k}) \text{ and } \mathbf{r} = 2\mathbf{i} - \mathbf{j} + 2\mathbf{k} + t(-\mathbf{i} - \mathbf{j} + \mathbf{k})$$

(c) Find the angle between the lines given in (b).

1 **r** = 5**i** − 2**j**

2 |**r**| = 13, 67.4° clockwise.

3 70.9°

4 (Answers are not unique, other answers are possible.)

(a) **r** = 3**i** + **k** + t(4**i** − **j** + 2**k**) (b) **r** = 4**i** − **j** + **k** + t(−7**i** + 3**j** − 3**k**)

5 (a) (1⅗, −1⅖) (b) Intersect at (7, 4, −3) (c) 121.5°

Sample questions and model answers

Core 4

1

(a) Express $f(x) = \dfrac{5x+1}{(x+2)(x-1)^2}$ in partial fractions.

(b) Hence show that $\displaystyle\int_{-1}^{0} f(x)\,dx = 1 - 2\ln 2$.

Remember the format when there is a repeated factor.

Take care with the denominator.

(a) Let $\dfrac{5x+1}{(x+2)(x-1)^2} \equiv \dfrac{A}{x+2} + \dfrac{B}{x-1} + \dfrac{C}{(x-1)^2}$

$\Rightarrow \dfrac{5x+1}{(x+2)(x-1)^2} \equiv \dfrac{A(x-1)^2 + B(x+2)(x-1) + C(x+2)}{(x+2)(x-1)^2}$

$\Rightarrow 5x+1 \equiv A(x-1)^2 + B(x+2)(x-1) + C(x+2)$

Putting $x = 1$ makes the factor $(x-1)$ equal to zero.

You could equate the x term or the constant term if you wish.

Let $x = 1$: $\qquad\qquad 6 = 3C \Rightarrow C = 2$

Let $x = -2$: $\qquad\qquad -9 = 9A \Rightarrow A = -1$

Equate x^2 terms: $\qquad 0 = A + B \Rightarrow B = 1$

$\therefore\ f(x) = -\dfrac{1}{x+2} + \dfrac{1}{x-1} + \dfrac{2}{(x-1)^2}$

(b) $\displaystyle\int_{-1}^{0} f(x)\,dx = \int_{-1}^{0}\left(-\dfrac{1}{x+2} + \dfrac{1}{x-1} + \dfrac{2}{(x-1)^2}\right)dx$

Remember the modulus sign.

$= \left[-\ln|x+2| + \ln|x-1| - 2(x-1)^{-1}\right]_{-1}^{0}$

$= -\ln 2 + \ln|-1| + 2 - (-\ln 1 + \ln|-2| + 1)$

$= -\ln 2 + 2 - \ln 2 - 1$

$\ln 1 = \ln|-1| = 0$
$\ln|-2| = \ln 2$.

$= 1 - 2\ln 2$

Core 4

2

Expand $(1 - 2x)^{-3}$ in ascending powers of x as far as the term in x^3 and state the set of values of x for which the expansion is valid.

Take care with the negatives.

Remember to include the restriction on x, even when it is not requested specifically in the question.

$(1-2x)^{-3} = 1 + (-3)(-2x) + \dfrac{(-3)(-4)}{2!}(-2x)^2 + \dfrac{(-3)(-4)(-5)}{3!}(-2x)^3 + \ldots$

$= 1 + 6x + 24x^2 + 80x^3 + \ldots$ provided $|2x| < 1$, i.e. $|x| < 0.5$

Sample questions and model answers (continued)

Core 4

3

Use the substitution $x = \tan u$ to show that $\displaystyle\int_0^1 \frac{1}{(1+x^2)^2}\,dx = \int_0^{\frac{\pi}{4}} \cos^2 u\,du$.

Hence show that $\displaystyle\int_0^1 \frac{1}{(1+x^2)^2}\,dx = \frac{\pi}{8} + \frac{1}{4}$.

> Show this essential side working.

Remember to change the x limits to u limits.

$$\int_0^1 \frac{1}{(1+x^2)^2}\,dx = \int_{x=0}^{x=1} \frac{1}{(1+x^2)^2} \frac{dx}{du}\,du \qquad \text{Let } x = \tan u$$

$$= \int_0^{\frac{\pi}{4}} \frac{1}{(\sec^2 u)^2} \sec^2 u\,du \qquad \Rightarrow \frac{dx}{du} = \sec^2 u$$

$$= \int_0^{\frac{\pi}{4}} \frac{1}{\sec^2 u}\,du \qquad 1 + x^2 = 1 + \tan^2 u$$

To integrate even powers of $\sin x$ or $\cos x$, change to double angles using $\cos 2x$ formula.

$$= \int_0^{\frac{\pi}{4}} \cos^2 u\,du \qquad = \sec^2 u$$

Limits:

When $x = 0$, $\tan u = 0 \Rightarrow u = 0$

When $x = 1$, $\tan u = 1 \Rightarrow u = \dfrac{\pi}{4}$

$$\int_0^{\frac{\pi}{4}} \cos^2 u\,du = \frac{1}{2}\int_0^{\frac{\pi}{4}} (1 + \cos 2u)\,du \qquad \cos 2u = 2\cos^2 u - 1$$

$$= \frac{1}{2}\left[u + \frac{1}{2}\sin 2u \right]_0^{\frac{\pi}{4}} \qquad \Rightarrow \cos^2 u = \tfrac{1}{2}(1 + \cos 2u)$$

$$= \frac{1}{2}\left(\frac{\pi}{4} + \frac{1}{2} - 0 \right)$$

$$= \frac{\pi}{8} + \frac{1}{4}$$

Core 3

4

(a) Solve the inequality $|2x - 3| < 11$

(b) Sketch the graph of $y = |f(x)|$ where $f(x) = (x-1)(x-2)(x-3)$

(a) $|2x - 3| < 11$

This first step is important.

$\Rightarrow -11 < 2x - 3 < 11$

You can simplify the inequality by doing the same thing to each part.

$\Rightarrow -8 < 2x < 14$ ← Add 3 to each part.

$\Rightarrow -4 < x < 7$. ← Divide each part by 2.

(b)

Parts of the graph of $y = f(x)$ that would be below the x-axis are reflected in the x-axis to give the graph of $y = |f(x)|$.

Sample questions and model answers (continued)

Core 3

5

The functions f and g are defined by:

$$f(x) = \sqrt{x-1} \quad x \geqslant 1$$

$$g(x) = 2x + 3 \quad x \in \mathbb{R}$$

(a) Find an expression for $f^{-1}(x)$ and state its domain and range.

(b) Find the value of:

(i) $fg(7)$

(ii) $gf(10)$

(iii) $f^{-1}g^{-1}(15)$

Set $y = f(x)$ and rearrange to make x the subject.

(a) $y = \sqrt{x-1}$

$\Rightarrow y^2 = x - 1$

$\Rightarrow x = y^2 + 1$

$\Rightarrow f^{-1}(x) = x^2 + 1$

The domain of f^{-1} is given by the range of f.

The domain of f^{-1} is $x \geqslant 0$ and the range of f^{-1} is $x \geqslant 1$.

The range of f^{-1} is given by the domain of f.

The order in which you apply the functions is important.

(b) (i) $fg(7) = f(17) = 4$

(ii) $gf(10) = g(3) = 9$

(iii) $f^{-1}g^{-1}(15) = f^{-1}(6) = 37$

Core 4

6

The equation of a straight line l is

$\mathbf{r} = 4\mathbf{i} - 2\mathbf{k} + t(3\mathbf{i} + \mathbf{k})$ where t is a scalar parameter.

Points A, B and C lie on the line l. Referred to the origin O, they have position vectors $\mathbf{a}$, $\mathbf{b}$ and $\mathbf{c}$ respectively.

(a) Find $\mathbf{a}$ and $\mathbf{b}$ and angle AOB, given that distance $OA = OB = 10$ units.

(b) Find $\mathbf{c}$, given that OC is perpendicular to l.

(a) $\underline{a} = (4 + 3t)\underline{i} + (-2 + t)\underline{k}$ for some value t. (1)

Since $|\underline{a}| = 10$,

Distance $OA = |\mathbf{a}|$.

$$\sqrt{(4 + 3t)^2 + (-2 + t)^2} = 10$$

$$16 + 24t + 9t^2 + 4 - 4t + t^2 = 100$$

$$10t^2 + 20t - 80 = 0$$

$$t^2 + 2t - 8 = 0$$

$$(t - 2)(t + 4) = 0$$

$$\Rightarrow t = 2 \text{ or } t = -4$$

Sample questions and model answers (continued)

This gives the position vectors of **a** and **b**.

Substituting for t in (1) gives

$\underline{a} = 10\underline{i}$ and $\underline{b} = -8\underline{i} - 6\underline{k}$

Let angle AOB be θ.

$|\underline{a}| = |\underline{b}| = 10$ and $\underline{a} \cdot \underline{b} = (10\underline{i}) \cdot (-8\underline{i} - 6\underline{k}) = -80$

$\underline{a} \cdot \underline{b} = |\underline{a}||\underline{b}| \cos \theta$

$\implies \cos \theta = \dfrac{-80}{100} = -0.8$

$\implies \theta = 143.1°$ (1 d.p.)

(b) $\underline{c} = (4 + 3t)\underline{i} + (-2 + t)\underline{k}$ for some value t.   (2)

3**i** + **k** is the direction vector of l and the scalar product of perpendicular vectors is zero.

Since $\underline{c}$ is perpendicular to l,

$((4 + 3t)\underline{i} + (-2 + t)\underline{k}) \cdot (3\underline{i} + \underline{k}) = 0$

$\implies 12 + 9t - 2 + t = 0$

$\implies t = -1$

Substituting for t in (2) gives $\underline{c} = \underline{i} - 3\underline{k}$.

Practice examination questions

Core 4

1 It is given that $f(x) = \dfrac{1}{(1+x)^2} + \sqrt{4+x}$.

Show that if x^3 and higher powers are ignored, $f(x) \approx a + bx + cx^2$ and find the values of a, b and c.

State the values of x for which the expansion is valid.

Core 4

2 (a) Find $\dfrac{dy}{dx}$ when $y = \cos^2 x \sin x$, expressing your answer in terms of $\cos x$.

(b) Using the substitution $u = \cos x$, or otherwise, find $\displaystyle\int \cos^2 x \sin x \, dx$.

Core 3

3 A curve has equation $y = \dfrac{x^2 - 4}{x + 1}$.

The equation of the normal to the curve at $(2, 0)$ is $ax + by + c = 0$, where a, b and c are integers.

Find the values of a, b, and c.

Core 3

4 The functions f and g are defined by:

$f{:}x \mapsto 5x - 7 \qquad x \in \mathbb{R}$

$g{:}x \mapsto (x + 1)(x - 1) \quad x \in \mathbb{R}$.

(a) Find the range of g.

(b) Find an expression for $fg(x)$.

(c) Determine the values of x for which $fg(x) = f(x)$.

Core 3

5 (a) Express $15 \sin x + 8 \cos x$ in the form $R \sin(x + \alpha)$ where $0 < \alpha < 90°$.

(b) State the maximum value of $15 \sin x + 8 \cos x$.

(c) Solve the equation $15 \sin x + 8 \cos x = 12$.

  Give all the solutions in the interval $0 < x < 360°$.

Core 4

6 Expressing y in terms of x, solve the differential equation

$$x \frac{dy}{dx} = y + yx,$$

given that $y = 4$ when $x = 2$.

Core 3

7 (a) Use the chain rule to differentiate:

  (i) $y = (4x - 3)^9$

  (ii) $y = \ln(5 - 2x)$

(b) Using your results from part (a):

  (i) Write down an expression for $\displaystyle\int (4x - 3)^8 \, dx$

  (ii) Calculate $\displaystyle\int_3^4 \frac{1}{5 - 2x} \, dx$

Practice examination questions (continued)

Core 4

8 (a) Use integration by parts to evaluate $\int_0^1 x\,e^{-2x}\,dx$.

(b) The diagram shows the graph of $y = x\,e^{-x}$.

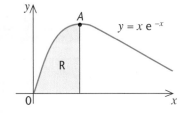

The curve has a maximum point at A.

The shaded region R is the area bounded by the curve, the x-axis and the vertical line through A.

(i) Show that the coordinates of A are $(1, e^{-1})$.
(ii) The region R is rotated completely about the x-axis.
    Calculate the volume of the solid generated.

Core 4

9 In a biology experiment, the growth of a population is being investigated.

A simple model is set up in which the rate of increase of the population at time t is proportional to the number, P, in the population at that time.

It is known that when $t = 0$, $P = 500$ and when $t = 10$, $P = 1000$.

Obtain the relationship between P and t and estimate the size of the population, according to the model, when $t = 20$.

Core 4

10 (a) Find the equation of the normal to the curve $x^3 + xy + y^2 = 7$ at the point $(1, 2)$.

(b) Find the coordinates of the two stationary points on the curve $x = t^2$, $y = t + \dfrac{1}{t}$.

Core 4

11 The region R, shaded in the diagram, is bounded by the curve $y = \sin 2x$ and the x-axis.

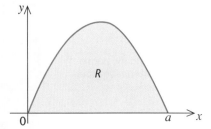

(a) Find the value of a.
(b) Show that the area of R is 1 square unit.
(c) Find the volume of the solid formed when R is rotated completely about the x-axis.

Core 4

12 (a) Find a vector equation of the line AB through $A(2, 2, 0)$ and $B(2, 3, 1)$.

(b) Show that the line AB and the line $\mathbf{r} = 2\mathbf{j} + \mathbf{k} + \lambda(\mathbf{i} + \mathbf{k})$, where λ is a scalar parameter, have no common point.

(c) Find the acute angle between the two lines.

Chapter 2

Probability and Statistics 2

The following topics are covered in this chapter:

- *Continuous random variables*
- *The Poisson distribution and approximations*
- *Estimation and sampling*
- *Hypothesis tests*

2.1 Continuous random variables

After studying this section you should be able to:

- *use the normal distribution model to calculate probabilities*
- *find probabilities and the mean and variance of a continuous variable with probability density function $f(x)$*

LEARNING SUMMARY

The normal distribution

The **normal variable** is a continuous random variable. Its probability function is represented by a bell-shaped curve, symmetric about the mean value μ.

The parameters of the normal distribution are μ and σ^2. If the continuous random variable X follows a normal distribution with these **parameters** then this is written as $X \sim N(\mu, \sigma^2)$. The standard deviation is σ.

> Tabulated values of probabilities for the normal distribution use Z as the variable.

The **standard normal variable** Z is used to calculate probabilities based on the Normal distribution where $Z = \dfrac{X - \mu}{\sigma}$ and $Z \sim N(0, 1)$.

Example
Find $P(15 < X < 17)$ where $X \sim N(15, 25)$

$$15 < X < 17 \Rightarrow \frac{15 - 15}{5} < Z < \frac{17 - 15}{5} \Rightarrow 0 < Z < 0.4$$

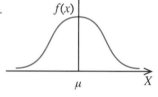

> Using the normal distribution tables.

$$P(15 < X < 17) = P(0 < Z < 0.4)$$
$$= \Phi(0.4) - \Phi(0) = 0.6554 - 0.5000$$
$$= 0.1554$$

You may need to use given information to set up and solve simultaneous equations to find the values of μ and σ.

Example
It is given that $P(X > 70) = 0.0436$ and $P(X < 45) = 0.3669$, where $X \sim N(\mu, \sigma^2)$. Find the values of μ and σ.

$$P(X > 70) = P\left(Z > \frac{70 - \mu}{\sigma}\right) = 0.0436$$

$$\Phi\left(\frac{70 - \mu}{\sigma}\right) = 1 - 0.0436 = 0.9564$$

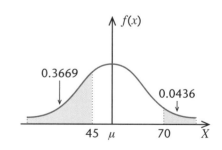

From the tables

$$\left(\frac{70-\mu}{\sigma}\right) = 1.71 \Rightarrow 70 - \mu = 1.71\sigma \qquad [1]$$

Similarly

$$P(X < 45) = P\left(Z < \frac{45-\mu}{\sigma}\right) = 0.3669$$

$$\Phi\left(\frac{45-\mu}{\sigma}\right) = 0.3669$$

In general,
$\Phi(-z) = 1 - \Phi(z)$.

$$\Phi\left(-\left(\frac{45-\mu}{\sigma}\right)\right) = 1 - 0.3669 = 0.6331$$

$$-\left(\frac{45-\mu}{\sigma}\right) = 0.34$$

$$45 - \mu = -0.34\sigma \qquad [2]$$

[1] − [2] gives $25 = 2.05\sigma \Rightarrow \sigma = 12.2$

Substituting in (1): $\qquad \mu = 49.1$

The probability density function, $f(x)$

The continuous random variable X is defined by its **probability density function** $f(x)$, together with the values for which it is valid.

Probabilities are given by areas under the curve.

$$P(c \leqslant X \leqslant d) = \int_c^d f(x)\mathrm{d}x$$

The total probability is 1.

If $f(x)$ is valid for $a \leqslant x \leqslant b$, then $\int_a^b f(x)\mathrm{d}x = 1$.

The probability of obtaining a specific value is zero.

$P(X = c) = 0$

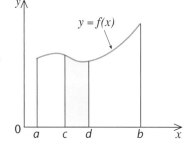

The mean is the **expectation** of X, written E(X). It is often denoted by μ.

If $f(x)$ is given in stages, then the integration is carried out in stages also.

> **KEY POINT**
>
> **Mean:** $\quad E(X) = \mu = \displaystyle\int_{\text{all } x} x f(x)\mathrm{d}x$
>
> **Variance:** $\quad \text{Var}(X) = \sigma^2 = E(X^2) - \mu^2$, where $E(X^2) = \displaystyle\int_{\text{all } x} x^2 f(x)\mathrm{d}x$

Example
The continuous random variable X has probability density function $f(x)$ given by $f(x) = \frac{1}{4}x$ for $1 \leqslant x \leqslant 3$. Find the mean μ, E(X^2) and the standard deviation σ.

$$\mu = E(X) = \int_{\text{all } x} x f(x)\mathrm{d}x = \int_1^3 \tfrac{1}{4}x^2\mathrm{d}x = \left[\frac{x^3}{12}\right]_1^3 = 2\tfrac{1}{6}$$

$$E(X^2) = \int_{\text{all } x} x^2 f(x)\mathrm{d}x = \int_1^3 \tfrac{1}{4}x^3\mathrm{d}x = \left[\frac{x^4}{16}\right]_1^3 = 5$$

$\sigma = \sqrt{\text{variance}}$

$$\text{Var}(X) = 5 - (2\tfrac{1}{6})^2 = \tfrac{11}{36} \Rightarrow \sigma = \sqrt{\tfrac{11}{36}} = 0.553 \text{ (3 d.p.)}$$

Progress check

1   Find $P(17 < X < 24)$ where $X \sim N(14, 16)$.

2   It is given that $X \sim N(\mu, \sigma^2)$ and $P(X < 30) = 0.7$
    Given also that $P(X > 35) = 0.1$, find μ and σ.

3   $f(x) = 1 - 0.5x,\ 0 \leqslant x \leqslant 2.$
    Find $E(X)$, $E(X^2)$ and $Var(X)$.

4   The continuous random variable X has probability density function $f(x) = 0.25$
    for $3 \leqslant x \leqslant a$ and $f(x) = 0$ otherwise.
    (a)   Find the value of a.
    (b)   Write down the value of the median of X.
    (c)   Calculate the standard deviation of X.

4   (a) 7   (b) 5   (c) 1.15 (3 s.f.)
3   $\frac{2}{3}, \frac{2}{3}, \frac{2}{9}$
2   $\mu = 26.5$
    $\sigma = 6.60$
1   0.2204

2.2 The Poisson distribution and approximations

After studying this section you should be able to:

- *understand the Poisson distribution*
- *use the Poisson approximation to the binomial distribution*
- *use the normal approximation to the binomial distribution*
- *use the normal approximation to the Poisson distribution*

LEARNING SUMMARY

The Poisson distribution

The **Poisson distribution** is used to model the number of occurrences, X, of an event in a given interval of space or time, when the events occur at a constant average rate, independently and at random.

If the mean number of occurrences in the interval is λ, then X ~ Po(λ).

> λ is the **parameter of** the distribution.

> Mean = variance

KEY POINT

If X ~ Po(λ):
$$P(X = x) = e^{-\lambda}\frac{\lambda^x}{x!} \qquad \text{for } x = 0, 1, 2, 3, \ldots$$

Mean, $\mu = \lambda$

Variance $= \lambda$ and standard deviation $= \sqrt{\lambda}$

> This is similar to using cumulative Binomial tables in Probability and Statistics 1.

For some values of λ you can use **cumulative Poisson probability tables** to find $P(X \leqslant x)$ and from this you can find other probabilities, for example:

$P(X < 2) = P(X \leqslant 1)$	0 1 2 3 ...
$P(X > 4) = 1 - P(X \leqslant 4)$	0 1 2 3 4 5 6 7 ...
$P(X \geqslant 3) = 1 - P(X \leqslant 2)$	0 1 2 3 4 5 ...
$P(X = 4) = P(X \leqslant 4) - P(X \leqslant 3)$	0 1 2 3 4 5 ...

Example

The number of letters received by the Adams family on a weekday may be modelled by a Poisson distribution with mean 3.75 and the number of letters received on a Saturday may be modelled by a Poisson distribution with mean 3. Determine the probability that:

(a) 4 letters are received on Monday

(b) more than 5 letters are received on Saturday.

> $\lambda = 3.75$ is not in the Poisson tables so you have to use the formula.

(a) X is the number of letters received on Monday.

$$X \sim Po(3.75) \text{ so } P(X = 4) = e^{-3.75}\frac{3.75^4}{4!} = 0.1937\ldots = 0.194 \text{ (3 s.f.)}$$

> $\lambda = 3$ is in the Poisson tables.

(b) Y is the number of letters received on Saturday.

$$Y \sim Po(3) \text{ so } P(Y > 5) = 1 - P(Y \leqslant 5) = 1 - 0.9161 = 0.0839$$

The Poisson approximation to the binomial distribution

The binomial distribution $X \sim B(n, p)$ is used to model the number of successes in n independent trials when the probability of success, p, is constant.

The mean of the binomial distribution is np and the variance is npq (where $q = 1 - p$).

With these conditions, $np \approx npq$. This fits in with the Poisson distribution where the mean and the variance are equal.

> **KEY POINT**
>
> When n is large ($n > 50$) and $np < 5$, $X \sim B(n, p)$ can be **approximated** by a Poisson distribution with the same mean, where $X \sim Po(np)$ approximately.

Example

The proportion of defective items produced by a machine is 2.5%. Find the probability of obtaining fewer than 3 defective items in a random sample of 100 items.

If X is the number of defective items in 100, then $X \sim B(100, 0.025)$.
The mean $np = 100 \times 0.025 = 2.5$
Since $n > 50$ and $np < 5$, use the Poisson approximation, $X \sim Po(2.5)$.

$$P(X = x) = \frac{\lambda^x}{x!}e^{-\lambda}$$

$$P(X < 3) = P(X = 0) + P(X = 1) + P(X = 2)$$

$$= e^{-2.5} + 2.5e^{-2.5} + \frac{2.5^2}{2!}e^{-2.5} = e^{-2.5}\left(1 + 2.5 + \frac{2.5^2}{2!}\right) = 0.54 \text{ (2 s.f.)}$$

It is often quicker to use cumulative probability tables, so make sure that you know how to use them.

If you have access to cumulative Poisson tables giving $P(X \leqslant r)$ for various values of r, then these tables can be used with $\lambda = 2.5$ and $P(X < 3) = P(X \leqslant 2) = 0.5438 = 0.54$ (2 s.f.).

The normal approximation to the binomial distribution

> **KEY POINT**
>
> When n is large enough to ensure that $np > 5$ and $nq > 5$ (where $q = 1 - p$), $X \sim B(n, p)$ can be **approximated** by a normal distribution with the same mean and the same variance, where $X \sim N(np, npq)$ approximately.

The larger the value of n and the closer that p is to 0.5, the better the approximation.

The normal distribution is continuous, whereas the binomial is discrete, so a **continuity correction** must be used.

Example

The probability that a person supports a particular organisation is 0.3. In a random sample of 50 people, find the probability that at least 20 support the organisation.

Always define the variable, giving its distribution if known.

If X is the number of people in 50 who support the organisation, then $X \sim B(50, 0.3)$.

$X \sim N(\mu, \sigma^2)$ with $\mu = 15$ and $\sigma^2 = 10.5 \Rightarrow \sigma = \sqrt{10.5}$

Mean $np = 50 \times 0.3 = 15$, variance $npq = 50 \times 0.3 \times 0.7 = 10.5$
Conditions for normal approximation: $np = 15 > 5$, $nq = 50 \times 0.7 = 35 > 5$

$\therefore$ $X \sim N(15, 10.5)$ approximately.

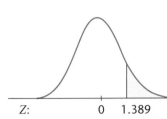

Z: 0   1.389

Applying the continuity correction,
$P(X \geqslant 20)$ becomes $P(X > 19.5)$.

$$P(X > 19.5) = P\left(Z > \frac{19.5 - 15}{\sqrt{10.5}}\right)$$

$$= P(Z > 1.389)$$

$$= 1 - \Phi(1.389) = 1 - 0.9176 = 0.082 \text{ (2 s.f.)}$$

Think of $X = 20$ as going from 19.5 to 20.5. Since 'at least 20' includes $X = 20$, find $P(X > 19.5)$ in the normal distribution.

Standardise
$$Z = \frac{X - \mu}{\sigma}$$
and then use standard normal tables.

The normal approximation to the Poisson distribution

The Poisson distribution, $X \sim \text{Po}(\lambda)$, is used to model the number of occurrences of an event, when events occur randomly. The mean is λ and the variance is λ.

> When λ is large ($\lambda > 15$ say), $X \sim \text{Po}(\lambda)$ can be approximated by a normal distribution with the same mean and variance, where $X \sim \text{N}(\lambda, \lambda)$ approximately.
>
> **KEY POINT**

Since the Poisson distribution is discrete, a continuity correction must be used.

Example

If $X \sim \text{Po}(20)$, find $P(15 < X < 22)$.

Since n is large, $X \sim \text{N}(20, 20)$ approximately.

You do not want to include 15 or 22 so go from 15.5 to 21.5

$P(15 < X < 22)$ becomes $P(15.5 < X < 21.5)$.

Apply the continuity correction.

Standardise the variables and then use normal tables.

$$P(15.5 < X < 21.5) = P\left(\frac{15.5 - 20}{\sqrt{20}} < Z < \frac{21.5 - 20}{\sqrt{20}}\right)$$

$$= P(-1.006 < Z < 0.335)$$

$$= \Phi(1.006) + \Phi(0.335) - 1$$

$$= 0.8427 + 0.6312 - 1$$

$$= 0.474 \text{ (3 s.f.)}$$

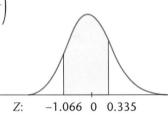

Z: −1.066 0 0.335

Progress check

1 The variable X follows a Poisson distribution with standard deviation 2.
   (a) Write down the value of λ.
   (b) Use cumulative probability tables to find
       (i) $P(X < 5)$ (ii) $P(X \geqslant 1)$ (iii) $P(X = 6)$

2 The number of telephone calls to a switchboard may be modelled by a Poisson distribution with mean 2.75 calls per 5-minute interval. Determine, to 3 significant figures, the probability that in a five-minute interval,
   (a)   there will be at most one call
   (b)   there will be exactly 5 calls.

3 $X \sim B(80, 0.06)$. Find, to 3 decimal places, $P(X = 4)$
   (a) by using the binomial distribution
   (b) by using a suitable approximation.

4 A newspaper reports that 62% of adults have an e-mail address. A random sample of 40 adults was selected. Using a suitable approximation, find the probability that at least 30 had an e-mail address.

5 Telephone calls reach a switchboard independently and at random at a rate of 30 per hour. Find the probability that in a randomly selected period of one hour there are fewer than 20 calls.

1 (a) 4 (b)(i) 0.6288 (ii) 0.9817 (iii) 0.1042
2 (a) 0.240 (3 s.f.) (b) 0.0838 (3 s.f.)
3 (a) 0.186 (b) 0.182
4 0.063 (2 s.f.)
5 0.028 (2 s.f.)

2.3 Estimation and sampling

After studying this section you should be able to:

- *distinguish between different sampling techniques*
- *find unbiased estimates for population parameters from a sample*
- *understand the theory relating to sampling distributions*
- *use the central limit theorem*

LEARNING SUMMARY

Sampling techniques

When you want information about a **population**, you could carry out a **census** of every member. The advantage is that you would have accurate and complete information. Disadvantages include cost and time requirements. Taking a census could also destroy the population. For example, if you were investigating the length of life of a calculator battery, carrying out a census would destroy the population.

More usually, a **sample** is taken. The sample should be representative of the whole population, so **bias** in the choice of sample members must be avoided.

In a **random sample** of size n, each member of the population has an equal chance of being selected. Also all subsets of the population of size n have an equal chance of constituting the sample.

Assigning a number to each member of a population and then drawing the numbers out of a hat is one method of obtaining a **simple random sample**. Another is to use **random number tables**, in which each digit has an equal chance of occurring.

If there is a periodic fault, then systematic sampling may fail to register it or may give it undue weight.

Random sampling from a very large population can be very laborious. It may be more convenient to carry out **systematic sampling**. This involves selecting every kth member of the population, for example, every 10th item from a particular machine on a production line.

The stratified sample can be constructed in proportion to the number of members in each stratum.

The method of **stratified sampling** is often used when the population is split into distinguishable layers or strata, such as students in each faculty in a college.

Unbiased estimates

When a **population parameter**, such as the mean or the variance, is **unknown**, then it is sensible to estimate it from a sample.

An **unbiased estimate** is one which, on the average, gives the true value, i.e. E(estimate) = true value of parameter.

The best unbiased estimate is the estimate with the smallest variance.

Key points from AS

- **Mean, variance and standard deviation**
  Revise AS pages 75–77

Make sure that you are familiar with the format given in your examination booklet and practise using it.

Best unbiased estimates:

For mean μ $\qquad \hat{\mu} = \bar{x} = \dfrac{\Sigma x}{n}$ $\qquad \bar{x}$ is sample mean

For variance σ^2 $\qquad \hat{\sigma}^2 = \dfrac{n}{n-1}\left(\dfrac{\Sigma x^2}{n} - \left(\dfrac{\Sigma x}{n}\right)^2\right)$

Sometimes S^2 is used for $\hat{\sigma}^2$

Alternative formats: $S^2 = \hat{\sigma}^2 = \dfrac{\Sigma(x-\bar{x})^2}{n-1}$, $S^2 = \hat{\sigma}^2 = \dfrac{1}{n-1}\left(\Sigma x^2 - \dfrac{(\Sigma x)^2}{n}\right)$.

> If you are given only summary data such as Σx or Σx^2, substitute into the appropriate formula.

A calculator in statistical mode can be used to obtain the value of $\hat{\sigma}$ directly from raw data. Look for the key marked $x\sigma n - 1$.

Sampling distributions

The most important sampling distribution is the distribution of the sample mean and you will need to be able to use it to find probabilities.

The sampling distribution of the mean

If all possible samples of size n are taken from $X \sim N(\mu, \sigma^2)$ and their sample means calculated, then these means form the sampling distribution of means, $\overline{X}$ which is also normally distributed.

> **KEY POINT**
>
> If $X \sim N(\mu, \sigma^2)$, then $\overline{X} \sim N\left(\mu, \dfrac{\sigma^2}{n}\right)$.

> The mean of $\overline{X}$ is the same as the mean of X.

The diagram shows the curves for X and for $\overline{X}$.

They are both symmetrical about μ.
The curve for $\overline{X}$ is much more squashed in, confirming the smaller standard deviation.

> The variance of $\overline{X}$ is much smaller than the variance of X.

> Standard deviation = $\sqrt{\text{variance}}$

The standard deviation of the sampling distribution, $\sigma/\sqrt{n}$, is known as the **standard error of the mean**.

The following result is extremely useful and should be learnt.

> **KEY POINT**
>
> If the distribution of X is not normal, then, by the **central limit theorem**, $\overline{X} \sim N\left(\mu, \dfrac{\sigma^2}{n}\right)$ approximately, **provided n is large**.

> $\hat{\sigma}^2$ is the unbiased estimate of σ^2.

If you do not know σ^2, the variance of X, then it can be estimated by $\hat{\sigma}^2$ and provided n is large, $\overline{X} \sim N\left(\mu, \dfrac{\hat{\sigma}^2}{n}\right)$ approximately.

Example

The heights of men in a particular area are normally distributed with mean 176 cm and standard deviation 7 cm. A random sample of 50 men is taken. Find the probability that the mean height of the men in the sample is less than 175 cm.

If X is the height, in centimetres, of a man from this area, then $X \sim N(176, 7^2)$.

For random samples of size 50, $\overline{X} \sim N\left(176, \dfrac{7^2}{50}\right)$, i.e. $\overline{X} \sim N(176, 0.98)$.

> In this example, since the population is normal, any size sample, large or small, could have been taken.

$$P(\overline{X} < 175) = P\left(Z < \frac{175 - 176}{\sqrt{0.98}}\right)$$
$$= P(Z < -1.010\ldots)$$
$$= 1 - \Phi(1.010)$$
$$= 1 - 0.8438$$
$$= 0.156 \text{ (3 d.p.)}$$

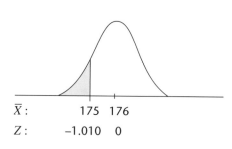

$\overline{X}:$ 175   176

$Z:$ −1.010   0

Progress check

1 The random variable X has mean μ and standard deviation σ.
Ten independent observations of X are
4.6, 3.9, 4.2, 5.8, 6.3, 4.2, 7.2, 6.0, 7.1, 5.4.

Find unbiased estimates of μ and σ.

2 A random sample of size 100 is taken from a normal distribution with mean
120 and standard deviation 5.
(a) Find $P(118.75 < \bar{X} < 120.5)$, where $\bar{X}$ is the sample mean.
(b) State, with a reason, whether your answer would be different if X is not
normally distributed.

1 5.47, 1.21
2 (a) 0.8351   (b) no, use central limit theorem since n is large

2.4 Hypothesis tests

After studying this section you should be able to:

- *understand the language used in hypothesis testing*
- *understand Type I and Type II errors*
- *understand how to obtain critical z-values*
- *perform the z-test for the mean*
- *perform tests for a binomial proportion (large and small sample sizes)*
- *perform a discrete test for a Poisson mean*

LEARNING SUMMARY

The language of hypothesis testing

When you want to know something about a population, you might perform a **hypothesis** or **significance test**.

A **null hypothesis**, H_0 is made about the population.
Then the **alternative hypothesis** H_1, the situation when H_0 is not true, is stated.

Example

When investigating the mean of a normal distribution, you might make the null hypothesis that the mean is 5. This is written H_0: $\mu = 5$.

The alternative hypothesis would be one of the following:

Use $\mu > 5$ if you suspect an *increase*.

Use $\mu \neq 5$ if you suspect a *change*.

- H_1: $\mu > 5$ (one-tailed, upper tail test)
- H_1: $\mu < 5$ (one-tailed, lower tail test)
- H_1: $\mu \neq 5$ (two-tailed test).

Use $\mu < 5$ if you suspect a *decrease*.

The formula for the test statistic depends on the sampling distribution. Examples are shown in the following text.

A **test statistic** is defined and its distribution when H_0 is true is stated.
Its value, based on information from a random sample taken from the population, is found. This is the **test value**.

The hypothesis test involves deciding whether or not the test value could have come from the distribution defined by the null hypothesis.
If the test value is in the main bulk of the distribution (the **acceptance region**) it is *likely* to have come from the distribution. If it is in the 'tail' end (the **rejection** or **critical region**), it is *unlikely* to have come from the distribution.

Probability theory is used to decide the placing of the boundary between 'likely' and 'unlikely'.

The decision rule (**rejection criterion**) is based on the **significance level** of the test, which is used to fix the boundaries for the rejection region. The boundaries are called **critical values**.

The levels used most often are 10%, 5% and 1%.

The significance level is a given as a percentage.

For **example**, for a significance level of 5%, the critical values are such that 5% of the distribution is in the critical (rejection) region,

There are two critical values in a two-tailed test.

i.e.   P(X > critical value) = 0.05, for a one-tailed, upper tail test
       P(X < critical value) = 0.05, for a one-tailed, lower tail test
       P(X > upper critical value) = 0.025, for a two-tailed test
       P(X < lower critical value) = 0.025, for a two-tailed test.

A test statistic is said to be significant if it lies in the critical region. If it lies in the acceptance region, then it is not significant.

Depending on the position of the test value, the **decision** is made.
If the test value lies in the critical region, H_0 is rejected in favour of H_1.
If the test value is in the acceptance region, H_0 is not rejected.

The **conclusion** is then stated in relation to the situation being tested.

Type I and Type II errors

In a hypothesis test, either H_0 is rejected or H_0 is not rejected.
There are occasions when the decision is incorrect and an error is made.

You make a Type I error when you reject a true null hypothesis.

If H_0 is rejected when it is in fact true, a **Type I** error is made.

The probability of making a Type I error is the same as the significance level of the test. For example, if the significance level is 5%, then P(Type I error) = 0.05

You make a Type II error when you accept a false null hypothesis.

If H_0 is accepted (i.e. not rejected) when it is in fact false, a **Type II** error is made.

To find the probability of making a Type II error, a specific value for H_1 must be given. Then P(Type II error) = P(H_0 is accepted when H_1 is true).

Critical values for z-tests

Some hypothesis tests involving the normal distribution are known as z-tests. In a z-test, the critical values for the rejection region can be found using the standard normal distribution, Z. The most commonly used values are summarised below.

5%
For one-tailed upper tail, $\Phi(z) = 0.95$
For two tailed, upper tail, $\Phi(z) = 0.975$
Use symmetry for lower tails.

10%
For one-tailed upper tail, $\Phi(z) = 0.90$
For two tailed, upper tail $\Phi(z) = 0.95$

1%
For one-tailed upper tail, $\Phi(z) = 0.99$
For two tailed, upper tail $\Phi(z) = 0.995$

Level	Type of test	Critical value	Decision rule
5%	One-tailed (upper)	1.645	Reject H_0 if $z > 1.645$
	One-tailed (lower)	−1.645	Reject H_0 if $z < -1.645$
	Two-tailed	±1.96	Reject H_0 if $z > 1.96$ or $z < -1.96$
10%	One-tailed (upper)	1.282	Reject H_0 if $z > 1.282$
	One-tailed (lower)	−1.282	Reject H_0 if $z < -1.282$
	Two-tailed	±1.645	Reject H_0 if $z > 1.645$ or $z < -1.645$
1%	One-tailed (upper)	2.326	Reject H_0 if $z > 2.326$
	One-tailed (lower)	−2.326	Reject H_0 if $z < -2.326$
	Two-tailed	±2.576	Reject H_0 if $z > 2.576$ or $z < -2.576$

z-test for the mean

This is used to test the mean μ of a population when you know the variance σ^2. If the population is **normal**, the samples can be **any size**, but if the population is **not normal**, then **large** samples must be taken, so that the central limit theorem can be applied.

The distribution considered for the test statistic is the sampling distribution of means (see page 72).

When the population is not normal, the central limit theorem is needed.

> Test statistic $Z = \dfrac{\bar{X} - \mu}{\sigma/\sqrt{n}} \sim N(0, 1).$
>
> **KEY POINT**

This formula can also be used if the variance of X is not known, provided n is large. In this case, the unbiased estimate $\hat{\sigma}^2$ is used for σ^2.

Example

A machine fills tins of baked beans such that the mass of a filled tin is normally distributed with mean mass 429 g and the standard deviation is 3 g. The machine breaks down and after being repaired, the mean mass of a random sample of 100 tins is found to be 428.45 g. Is this evidence, at the 5% level, that the mean mass of tins filled by this machine has decreased? Assume that the standard deviation remains unaltered.

Define your variables.

Let X be the mass, in grams, of a filled tin and let the mean after the repair be μ.

State the hypotheses.

H_0: $\mu = 429$; H_1: $\mu < 429$.

State the distribution of the test statistic assuming the null hypothesis is true.

Consider the sampling distribution of means $\bar{X} \sim N\left(429, \dfrac{3^2}{100}\right)$.

State the type of test and the rejection criterion.

Perform a one-tailed (lower tail) test at the 5% level and reject H_0 if $z < -1.645$

Calculate the value of the test statistic.

$\bar{x} = 428.45$, so

$$z = \frac{\bar{x} - \mu}{\sigma/\sqrt{n}} = \frac{428.45 - 429}{3/\sqrt{100}} = -1.833\ldots$$

0.05

$\bar{X}:$ 429

$Z:$ -1.645 0

Decide whether to reject H_0 or not and relate your conclusion to the question.

Since $z < -1.645$, the test value lies in the critical region and H_0 is rejected.

There is evidence, at the 5% level, that the **mean** mass of a filled tin has decreased.

Alternative method

Alternatively, a **probability method** can be used.
The rejection rule is to reject H_0 if $P(\bar{X} < 428.45) < 0.05$

$$P(\bar{X} < 428.45) = P\left(Z < \frac{428.45 - 429}{3/\sqrt{100}}\right)$$

$$= P(Z < -1.833\ldots) = 0.0334$$

0.0334

$\bar{X}:$ 428.45   429

Since $P(\bar{X} < 428.45) < 0.05$, H_0 is rejected and the conclusion is as above.

Tests for a binomial proportion

This binomial test is used to test the proportion of successes, p, in a binomial population $X \sim B(n, p)$.

Large sample size

Normal approximation to the binomial, see page 68.

When n is large (such that $np > 5$ and $nq > 5$) a normal approximation to the binomial distribution can be used.

When testing in the upper tail, subtract 0.5, and when testing in the lower tail, add 0.5.

A continuity correction of $+0.5$ or -0.5 is needed as the binomial distribution is discrete and the normal distribution is continuous.

> Test statistic $Z = \dfrac{(X \pm 0.5) - np}{\sqrt{npq}} \sim N(0, 1)$, provided n is large.
>
> **KEY POINT**

The hypothesis test follows the same pattern as the z-test for the mean.

Example

A coin is tossed 50 times and 30 heads are obtained. Is this evidence, at the 5% level of significance, that the coin is biased in favour of heads?

Define the variable.

Let the probability that the coin shows heads be p.
If X is the number of heads in 50 tosses, then $X \sim B(50, p)$.

State the hypotheses and also the distribution if the null hypothesis is true.

H_0: $p = 0.5$ (the coin is fair), H_1: $p > 0.5$ (the coin is biased in favour of heads). If the null hypothesis is true, then $X \sim B(50, 0.5)$.

Justify the normal approximation.

But n is large such that $np = 50 \times 0.5 = 25 > 5$ and $nq = 25 > 5$, so $X \sim N(np, npq)$ approximately. Since $npq = 50 \times 0.5 \times 0.5 = 12.5$, $X \sim N(25, 12.5)$.

State the rejection criterion.

At the 5% level, reject H_0 if $z > 1.645$

29.5 is used because you need to test whether the whole rectangle representing $x = 30$ (from 29.5 to 30.5) is in the upper tail critical region.

The test value is $x = 30$, so using the continuity correction,
$$z = \frac{(x - 0.5) - np}{\sqrt{npq}} = \frac{29.5 - 25}{\sqrt{12.5}} = 1.27 \ldots$$

Since $z < 1.645$, the test value does not lie in the critical region and H_0 is not rejected.

Relate the conclusion to the question.

At the 5% level there is not sufficient evidence to say that the coin is biased in favour of heads.

Binomial test when the sample size is small

When n is small, the normal approximation does not apply, so binomial probabilities are used to decide whether the test value is in the critical region or not.

For **example**, at the 5% level, the decision rules are:

- for a one-tailed (upper tail) test, reject H_0 if $P(X \geqslant \text{test value}) < 0.05$
- for a one-tailed (lower tail) test, reject H_0 if $P(X \leqslant \text{test value}) < 0.05$
- for a two-tailed test, reject H_0 if $P(X \leqslant \text{test value}) < 0.025$ or $P(X \geqslant \text{test value}) < 0.025$

Example

When a die was thrown 10 times, a six occurred 4 times. Is this evidence, at the 10% level of significance, that the die is biased in favour of sixes?

Let the probability of obtaining a six be p.

This is a small sample, so the normal approximation cannot be used.

If X is the number of sixes in 10 throws, then $X \sim B(10, p)$.
H_0: $p = \frac{1}{6}$ (the die is fair), H_1: $p > \frac{1}{6}$ (the die is biased in favour of sixes).

This is a one-tailed (upper tail) test.

If the null hypothesis is true, then $X \sim B(10, \frac{1}{6})$.
Reject H_0 if the test value of $x = 4$ lies in the critical region (the upper tail 10%)
i.e. reject H_0 if $P(X \geqslant 4) < 0.1$

It may be possible to calculate the probability using cumulative binomial tables. Check in your examination booklet.

$$P(X \geqslant 4) = 1 - P(X < 4)$$
$$= 1 - \left((\tfrac{5}{6})^{10} + {}^{10}C_1(\tfrac{5}{6})^9(\tfrac{1}{6}) + {}^{10}C_2(\tfrac{5}{6})^8(\tfrac{1}{6})^2 + {}^{10}C_3(\tfrac{5}{6})^7(\tfrac{1}{6})^3\right)$$
$$= 1 - 0.930 \ldots = 0.070 \; (2 \text{ s.f.})$$

Since $P(X \geqslant 4) < 0.1$, H_0 is rejected.

There is evidence at the 10% level, that the die is biased in favour of sixes.

Test for a Poisson mean

The test statistic is X, the number of occurrences, where $X \sim Po(\lambda)$.

When λ is **small**, the test is similar to the small sample binomial test.

You will not be required to carry out this test.

When λ is **large**, $X \sim N(\lambda, \lambda)$ approximately and the test is similar to the large sample binomial test. A continuity correction is needed.

Example

The office manager claims that the average number of telephone calls per minute received by her office switchboard is 4.5. Her supervisor maintains that it is less than 4.5. In a randomly selected minute, 2 calls were received. A hypothesis test is conducted at the 5% level. Whose claim is upheld?

Let X be the number of calls per minute. Assuming that calls occur randomly and independently, $X \sim \text{Po}(\lambda)$.

> This is a one-tailed (lower tail) test.

$H_0: \lambda = 4.5$, $H_1: \lambda < 4.5$. If the null hypothesis is true, $X \sim \text{Po}(4.5)$.

The test value is $x = 2$, so, at the 5% level, reject H_0 if $P(X \leqslant 2) < 0.05$

> Since λ is small, the method is similar to the small sample binomial test.

$$P(X \leqslant 2) = e^{-4.5} + 4.5e^{-4.5} + \frac{4.5^2}{2!}e^{-4.5} = e^{-4.5}\left(1 + 4.5 + \frac{4.5^2}{2!}\right) = 0.173 \ldots$$

Since $P(X \leqslant 2) > 0.05$, the test value does not lie in the critical region and H_0 is not rejected. The office manager's claim that the mean is 4.5 is upheld.

Progress check

1 The variable X is modelled by a normal distribution with mean μ and standard deviation 2. The mean of a random sample of 10 items from X is 20.9. Test, at the 5% level, the hypotheses $H_0: \mu = 20$, $H_1: \mu > 20$.

2 The heights of a particular plant have mean μ and variance σ^2 (both unknown). The heights H, in centimetres, of a random sample of 100 plants are summarised as follows: $\Sigma h = 2941$, $\Sigma h^2 = 88\,502$.

   (a) Find unbiased estimates for μ and σ.
   (b) Does the sample data support the claim that the mean height is less than 30 cm? Perform a hypothesis test at the 10% level and state your hypotheses.

> Hint:
> In (a) use small sample binomial test.
> In (b) use large sample binomial test.

3 A random observation, x, is taken from $X \sim \text{B}(n,p)$.
Test, at the 5% level, the null hypothesis that $p = 0.4$, against the alternative hypothesis that $p < 0.4$
   (a) if $n = 10$ and $x = 1$,   (b) if $n = 100$ and $x = 32$.

4 A single observation is taken from a Poisson distribution with mean λ and used to test the null hypothesis $\lambda = 7$ against the alternative hypothesis $\lambda \neq 7$. What is the conclusion if the hypothesis test is carried out at the 10% level and the observation is

> Cumulative Poisson tables are needed for (b).

   (a) 3   (b) 13?

1 $z = 1.423$, do not reject H_0.
2 (a) 29.41, 4.503   (b) $H_0: \mu = 30$, $H_1: \mu > 30$; $z = -1.310$, lies in critical region so reject H_0 and uphold claim.
3 (a) $P(X \leqslant 1) = 0.0464 < 0.05$, reject H_0, conclude $p < 0.4$
   (b) $z = -1.531$, $z > -1.645$, do not reject H_0, conclude that p could be 0.4
4 (a) $P(X < 3) > 0.05$, H_0 is not rejected.   (b) $P(X > 13) < 0.05$, H_0 is rejected.

Sample questions and model answers

1

The continuous random variable X has probability density function $f(x)$ given by

$$f(x) = \begin{cases} kx(2-x) & 0 \leqslant x \leqslant 2 \\ 0 & \text{otherwise} \end{cases}$$

where k is a constant.

(a) Show that $k = \frac{3}{4}$ and find the exact value of $P(X > 1\frac{1}{2})$.

(b) Find $E(X)$ and $E(X^2)$.

(c) Show that the standard deviation, σ, is 0.447 correct to three decimal places.

The total area under the curve is 1.

As k is a constant, it can be taken outside the integration.

(a)
$$\int_0^2 f(x)dx = 1$$

$$\implies k\int_0^2 (2x - x^2)dx = 1$$

$$\implies k\left[x^2 - \frac{x^3}{3}\right]_0^2 = 1$$

$$\implies k(4 - \tfrac{8}{3} - 0) = 1$$

$$\implies k = \tfrac{3}{4}$$

$$f(x) = \tfrac{3}{4}(2x - x^2)$$

$$P(X > 1\tfrac{1}{2}) = \int_{1\frac{1}{2}}^2 f(x)dx$$

$$= \tfrac{3}{4}\int_{1\frac{1}{2}}^2 (2x - x^2)dx$$

Substitute the limits carefully, working in fractions to obtain the exact answer.

$$= \tfrac{3}{4}\left[x^2 - \frac{x^3}{3}\right]_{1\frac{1}{2}}^2$$

$$= \tfrac{3}{4}(4 - \tfrac{8}{3} - (\tfrac{9}{4} - \tfrac{9}{8}))$$

$$= \tfrac{5}{32}$$

If you do not spot that $f(x)$ is symmetrical about $x = 1$, find $E(X)$ by integration.

(b) By symmetry, $E(X) = 1$

$$E(X^2) = \int_0^2 x^2 f(x)dx$$

$$= \tfrac{3}{4}\int_0^2 (2x^3 - x^4)dx$$

$$= \tfrac{3}{4}\left[\frac{2x^4}{4} - \frac{x^5}{5}\right]_0^2$$

$$= \tfrac{3}{4}(8 - \tfrac{32}{5})$$

$$= 1.2$$

(c) $\text{Var}(X) = E(X^2) - [E(X)]^2$

$[E(X)]^2$ is often written $E^2(X)$.

$$= 1.2 - 1$$

$$= 0.2$$

$$\sigma = \sqrt{0.2} = 0.447 \text{ (3 d.p.)}$$

Sample questions and model answers *(continued)*

2

In an advanced level examination taken by a large number of candidates, the marks were distributed normally with mean mark 68.7 and standard deviation 5.4.

A random sample of 100 scripts is taken and their mean mark denoted by $\bar{X}$.

(a) State the distribution of $\bar{X}$.

(b) Find the probability that the mean mark of the 100 scripts is between 68 and 70, giving your answer correct to 2 significant figures.

Let X be the examination mark, so $X \sim N(68.7, 5.4^2)$

(a) For samples of size 100,

$$\bar{X} \sim N\left(68.7, \frac{5.4^2}{100}\right), \text{ i.e. } \bar{X} \sim N(68.7, 0.2916)$$

(b) $P(68 < \bar{X} < 70) = P\left(\dfrac{68 - 68.7}{\sqrt{0.2916}} < Z < \dfrac{70 - 68.7}{\sqrt{0.2916}}\right)$

$\qquad\qquad\qquad\quad = P(-1.296 < Z < 2.407)$

$\qquad\qquad\qquad\quad = \Phi(1.296) + \Phi(2.407) - 1$

$\qquad\qquad\qquad\quad = 0.9026 + 0.9919 - 1 = 0.89 \text{ (2 s.f.)}$

If the degree of approximation is not specified in the question or your examination rubric, then approximate sensibly.

Sample questions and model answers *(continued)*

3

The lengths of a population of snakes are normally distributed with standard deviation 8 cm and unknown mean μ cm.

A random sample of 10 snakes is taken and their mean length calculated.

When a significance test, at the 10% level, is performed, the hypothesis that μ is 37.5 is rejected in favour of the hypothesis that it is greater than 37.5 cm.

(a) What can be said about the value of the sample mean?

(b) Explain briefly what is meant, in the context of this question, by a Type I error, and state the probability of making a Type I error.

(c) Find the probability of making a Type II error when $\mu = 41.1$

(a) If X is the length, in centimetres, then $X \sim N(\mu, 8^2)$.

H_0: $\mu = 37.5$

H_1: $\mu > 37.5$

For the significance test, the sampling distribution of means is considered.

If $\mu = 37.5$, then $\bar{X} \sim N\left(37.5, \dfrac{8^2}{10}\right)$.

From standard normal tables, $\Phi(z) = 0.9 \Rightarrow z = 1.282$

The critical value for a one-tailed (upper tail) at the 10% level is 1.282

H_0 is rejected if z is in the upper tail 10% of the distribution.

Since H_0 is rejected, $z > 1.282$,

Test statistic is $Z = \dfrac{\bar{X} - \mu}{\sigma/\sqrt{n}}$.

i.e. $\dfrac{\bar{x} - 37.5}{8/\sqrt{10}} > 1.282$

$\bar{x} > 40.74$

	Z:	0	1.282
	$\bar{X}$	37.5	40.74

10%

(b) A Type I error is made when the null hypothesis, $\mu = 37.5$, is rejected, when the mean population length is in fact 37.5

The probability of making a Type I error is the same as the level of significance of the test.

P(Type I error) = 10%

(c) H_0: $\mu = 37.5$

H_1: $\mu = 41.1$

P(Type II error) = P(H_0 is accepted | H_1 is true)

From (a), H_0 is accepted if $\bar{x} < 40.74$

If H_1 is true, $\bar{X} \sim N\left(41.1, \dfrac{8^2}{10}\right)$

P(Type II error) = $P\left(\bar{X} < 40.74 \mid \bar{X} \sim N\left(41.1, \dfrac{8^2}{10}\right)\right)$

$= P\left(Z < \dfrac{40.74 - 41.1}{8/\sqrt{10}}\right)$

$= P(Z < -0.142)$

$= 1 - 0.5565$

$= 0.44$ (2 s.f.)

P (Type II error)

	$\bar{X}$:	40.74 41.1

Practice examination questions

1 The variable X is normally distributed with mean μ and variance 16.
The null hypothesis $\mu = 20$ is to be tested against the alternative hypothesis $\mu \neq 20$.

The mean of a random sample of 10 observations from X is 17.2.

Test at the 5% level whether this provides evidence that μ is not 20.

2 The probability that a person suffers from a particular health condition is 0.01.

(a) Using a suitable approximation, find the probability that in a randomly chosen sample of 80 people, fewer than 3 suffer from the health condition.

(b) Find the minimum sample size in order that the probability of including at least one with the health condition is greater than 95%.

3 The continuous random variable X has probability density function $y = k$, where k is a constant, for $2 \leqslant x \leqslant 12$ and $f(x) = 0$ otherwise.

(a) Find the value of k.

(b) Calculate the probability that X lies within one standard deviation of the mean, giving your answer correct to 2 decimal places.

4 The volume, X litres, of paint in a tub may be modelled by a normal distribution with mean 3.5 and standard deviation σ.

(a) Assuming that $\sigma = 0.8$, find the value of $P(X < 3.3)$.
(b) Find the value of σ so that 95% of tubs contain more than 3.25 litres of paint.

5 The random variable X has a binomial distribution with $n = 10$ and $p = 0.4$.
The mean of 60 random observations of X is $\bar{X}$.

Find $P(\bar{X} < 3.5)$, explaining what part the central limit theorem has played in your answer.

Practice examination questions *(continued)*

6 A beetle infestation is discovered in the trees in a small copse. If more than 35% of the trees are infected it is impractical to treat the trees with chemicals and they will have to be felled.

BINOMIAL TEST
– small sample

(a) The representative from the ministry checked a random sample of 10 trees and found that 6 were infected. Stating any assumptions, use a hypothesis test, at the 10% significance level, to decide whether the infestation should be treated with chemicals or the trees in the copse felled.

BINOMIAL TEST
– large sample

(b) The estate manager checked a random sample of 30 trees and found that 14 were infected. Using a suitable approximation, carry out a significance test, again at the 10% level. On the basis of this sample, how should the infestation be dealt with?

7 The continuous random variable X has probability function $f(x)$ where $f(x) = k(x - x^3)$, $0 \leqslant x \leqslant 1$ and $f(x) = 0$ otherwise.

(a) Show that $k = 4$.

(b) Find $E(X)$.

(c) Find $P(X < 0.5)$.

(d) Find $Var(X)$.

8 A village newsagent finds that the demand for the local newspaper on a Monday follows a Poisson distribution with mean 16.

(a) Find the probability that there are fewer than 16 requests for the newspaper on a Monday.

(b) How many copies of the newspaper should there be in stock so that the demand is not satisfied on fewer than 5% of Mondays?

Chapter 3
Mechanics 2

The following topics are covered in this chapter:

- Projectiles
- Equilibrium of a rigid body
- Centre of mass

- Collisions and impulse
- Uniform circular motion
- Work, energy and power

3.1 Projectiles

After studying this section you should be able to:

- model the motion of a projectile moving under constant acceleration
- understand the limitations of the model
- solve problems by considering, separately, the vertical and horizontal components of velocity
- derive and apply the trajectory formula

LEARNING SUMMARY

Projectiles

In the standard model for projectiles, air resistance is ignored and the acceleration due to gravity g m s^{-2} is taken to be constant. According to this model:

> - The only force acting on a projectile, once in flight, is its weight.
> - The horizontal component of velocity is constant.
> - The vertical component of velocity is subject to a constant acceleration g m s^{-2}.
>
> **KEY POINT**

The approach used for many projectile problems is to resolve the initial velocity into horizontal and vertical components and to treat these two parts separately.

Example

A stone is thrown horizontally with a speed of 10 m s^{-1} from the top of a cliff. The cliff is 98 m high and the stone lands in the sea d m from its base. Take $g = 9.8$ m s^{-2} and show that $d = 20\sqrt{5}$.

You can assume that a cliff is vertical.

Choosing downwards to be positive.

Vertically: $u = 0$

$\qquad\qquad a = 9.8$

$\qquad\qquad s = 98.$

Using $s = ut + \frac{1}{2}at^2$

Even a very simple diagram can help you to visualise the problem.

gives $98 = 4.9t^2$

$\qquad \Rightarrow t = \sqrt{20} = 2\sqrt{5}.$

Horizontally: $d = 10 \times 2\sqrt{5}$

$\qquad\qquad \Rightarrow d = 20\sqrt{5}.$

The initial velocity of the projectile may be at angle θ to the horizontal.

Example

A ball is struck with velocity 20 m s^{-1} at 40° to the horizontal from a point 1 m above the ground. Find the maximum height reached by the ball. Take $g = 9.8$ m s^{-2}.

Choosing upwards to be positive.

Vertically: $u = 20 \sin 40°$
              $a = -9.8$

At max ht: $v = 0$.

Using $v = u + at$
              $0 = 20 \sin 40° - 9.8t$
              $\Rightarrow t = 1.3118 \ldots .$

Using $s = ut + \frac{1}{2}at^2$.

At max ht $s = 20 \sin 40° \times 1.3118 - 4.9 \times 1.3118^2$
              $= 8.432 \ldots .$

The maximum height is 9.43 m to 2 d.p.

The model

In the usual **particle** model for projectiles, air resistance is ignored and the only force taken to act on the particle is its weight which is constant.
According to this model:

Check the value of g used on your exam paper. Make sure that you use the given value or you may lose marks.

- The horizontal component of velocity remains constant throughout the motion.
- The vertical component of velocity is subject to a constant downward acceleration of g ms^{-2}.

The accuracy of the model, in predicting the motion of a projectile, depends on the extent to which the initial assumptions are satisfied.

For example, the model will provide accurate information about a projectile such as a stone or a dart over a short distance, but will give poor results for light objects that are affected more by air resistance.

Be careful about using formulae for time of flight, greatest height, range etc. without justification or, again, marks may be lost.

When using the model, a first step is usually to resolve the velocity of a projectile into horizontal and vertical components. The two parts are then treated separately.

A particle moving with speed v at angle θ to the horizontal has vertical component $v \sin \theta$ and horizontal component $v \cos \theta$.

The velocity is usually shown on a diagram as in Figure 1 but is used in equations as in Figure 2.

If you know the horizontal and vertical components of a velocity then you can combine them to find its magnitude (speed) and direction.

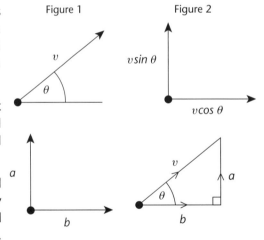

Figure 1   Figure 2

$$v = \sqrt{a^2 + b^2} \qquad \tan \theta = \frac{a}{b}$$

Upwards is taken to be positive so the acceleration is shown as negative.

Example
A stone is thrown with speed 15 ms^{-1} at an angle of 60° above the horizontal. Find its speed and direction after 1 second. Take $g = 9.8$ ms^{-2}.

Vertically: using $v = u + at$

$$v_a = 15 \sin 60° - 9.8$$

$$= 3.19038 \ldots$$

Horizontally: $v_b = 15 \cos 60°$
$$= 7.5$$
$$v = \sqrt{7.5^2 + 3.19038^2} = 8.15 \text{ to 3 s.f.}$$

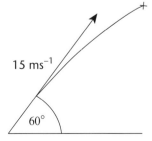

15 ms^{-1}

60°

$$\tan \theta = \frac{3.19038}{7.5} \Rightarrow \theta = 23.0° \text{ to 1 d.p.}$$

After 1 second the stone has speed 8.15 ms^{-1} at 23° above the horizontal.

The trajectory formula

Taking the point of projection as the origin, the position of a projectile may be described in terms of (x, y) coordinates.

Horizontally: $x = v \cos \theta \times t \Rightarrow t = \dfrac{x}{v \cos \theta}$ (1)

Vertically: $y = v \sin \theta \times t - \frac{1}{2}gt^2$ (2)

Substituting for t in (2) gives: $y = v \sin \theta \times \dfrac{x}{v \cos \theta} - \dfrac{gx^2}{2v^2 \cos^2 \theta}$

This is the Cartesian equation of the path of the projectile. It is known as the trajectory formula.

so: $\quad y = x \tan \theta - \dfrac{gx^2}{2v^2}(1 + \tan^2 \theta)$

Note that:
$$\dfrac{1}{\cos^2 \theta} \equiv \sec^2 \theta \equiv (1 + \tan^2 \theta).$$

Example

A ball is kicked with speed 20 ms^{-1} at 30° above the horizontal towards a wall 5 m high. The wall is 15 m from the point where the ball is kicked. Take $g = 9.8$ ms^{-2}. Will the ball hit the wall or clear it? Consider the effect of any assumptions.

Using the formula: $\quad y = x \tan \theta - \dfrac{gx^2}{2v^2}(1 + \tan^2 \theta)$

$$y = 15 \tan 30° - \frac{9.8 \times 15^2}{2 \times 20^2}(1 + \tan^2 30°)$$

$$= 4.99 \text{ m to 3 s.f.}$$

According to the model, the ball will hit the wall close to the top.
If the size of the ball is taken into account along with the effect of air resistance then it is clear that the ball cannot clear the wall.

Progress check

1 A small object is projected from ground level with speed 30 ms^{-1} at 40° above the horizontal. Take $g = 10$ ms^{-2}.

  (a) Find the speed and direction of the object after 1.5 seconds.

  (b) The object just clears a tree 25 m from the point of projection. How high is the tree?

1 (a) 23.4 ms^{-1}, 10.6° above horiz (b) 15.2 m

3.2 Equilibrium of a rigid body

After studying this section you should be able to:

- find the moment of a force
- understand the conditions for equilibrium of a rigid body under the action of coplanar forces
- solve problems involving coplanar forces

LEARNING SUMMARY

Moments

The **moment of a force** about a point is a measure of the turning effect of the force about the point.

It is found by multiplying the force by its perpendicular distance from the point.

The moment of a force acts either clockwise ↻ or anticlockwise ↺ about the point.

The moment of F about
P is Fd N m ↺ .

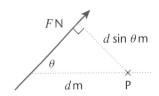

The moment of F about P
is $Fd \sin \theta$ N m ↺ .

> The units of force (N) are multiplied by the units of distance (m) to give N m.

When there is more than one force, the **resultant moment** about a point is found by adding the separate moments. One direction is taken to be positive and the other negative.

Example Find the resultant moment about O of the forces shown in the diagram.

Taking ↺ as positive:
The total moment is $8 \times 3 - 10 \times 2$ N m ↺

$$= 4 \text{ N m} \ ↺ .$$

Equilibrium

For an object to be in **equilibrium**, the resultant force acting on it must be zero and the resultant moment must be zero.

Example
The diagram shows a light rod AB in equilibrium. Find the values of F and d.

> In this context, a light rod is a rod is taken to have a mass that is so small that it can be ignored in any calculations.

Resolving vertically $F - 30 - 20 = 0$

$\Rightarrow \qquad\qquad\qquad\qquad\qquad F = 50$

Taking moments about C (written as M(C)) will give an equation involving d.

M(C) gives $\qquad\quad 30d - 20 \times 1.5 = 0$

$\Rightarrow \qquad\qquad\qquad\qquad 30d = 30$

$\Rightarrow \qquad\qquad\qquad\qquad\quad d = 1.$

Forces that act in the same plane are coplanar.

You may need to show that a rigid body is in equilibrium under a system of **coplanar** forces *or* to *use* the fact that a rigid body is in equilibrium to find something about the forces acting on it.

To show that a rigid body is in equilibrium under the action of coplanar forces you need to establish that:

- the vector sum of the forces is zero
- the sum of the moments about some point is zero.

You only need to establish this for *one* point and you can choose any point that is convenient.

If a body is in equilibrium under some coplanar forces then you know that:

- the vector sum of the forces will be zero
- the sum of the moments about any point you choose will be zero.

> The conditions for equilibrium can be used to produce equations. The equations can then be solved to find the unknown values in a problem.

KEY POINT

Example

A uniform plank AB of length 5 m and mass 20 kg rests horizontally on two supports, one at each end. A mass of 10 kg is positioned on the plank 1 m from B. Find the reaction at each of the supports. Take $g = 9.8 \text{ ms}^{-2}$.

The first step is to draw a clearly labelled diagram.

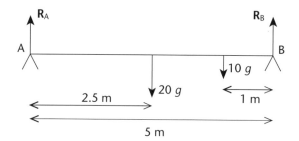

The plank is uniform so its weight acts at its mid-point.

The vector sum of the forces is zero so the total upward force must equal the total downward force.

Vertically:
$$\mathbf{R}_A + \mathbf{R}_B = 20g + 10g$$
$$= 30g$$

M(A) gives:
$$5\mathbf{R}_B = 20g \times 2.5 + 10g \times 4$$
$$= 90g$$
$$\mathbf{R}_B = 18g$$
$$= 176.4 \text{ N (reaction at B)}$$

Substitute for $\mathbf{R}_B$:
$$\mathbf{R}_A + 18g = 30g$$
$$\mathbf{R}_A = 12g$$
$$= 117.6 \text{ N (reaction at A)}.$$

Leaning ladders

The situation where a ladder leans against a wall, possibly supporting a load at some point along its length, provides a rich source of questions on equilibrium. It requires forces and moments to be considered and includes the theory of friction.

Example

A uniform ladder of mass m kg rests against a smooth vertical wall with its lower end on rough horizontal ground. A man of mass $4m$ kg has climbed three-quarters of the way up the ladder but cannot go any higher or it will slip. The ladder makes an angle of $60°$ with the horizontal.
Find the coefficient of friction between the ladder and the ground.

Start by drawing a clearly labelled diagram.

The wall is smooth so there is no vertical force at the top of the ladder.

The length of the ladder has been written as $4l$ to simplify taking moments.

Friction acts at the bottom of the ladder to oppose the tendency for it to slip.

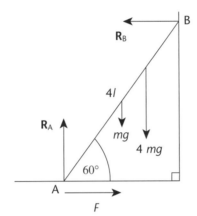

In this example, three equations may be formed by resolving the forces in two directions and taking moments.

The equations provide enough information to solve the problem.

Vertically:
$$\mathbf{R}_A = mg + 4mg \tag{1}$$
$$= 5mg$$

Horizontally:
$$F = \mathbf{R}_B \tag{2}$$

M(A):
$$\mathbf{R}_B \times 4l \sin 60° = mg \times 2l \cos 60° + 4mg \times 3l \cos 60°$$

giving:
$$\mathbf{R}_B \times 2\sqrt{3}l = mgl + 6mgl$$

so:
$$\mathbf{R}_B = \frac{7mg}{2\sqrt{3}} \tag{3}$$

From (2) and (3)
$$F = \frac{7mg}{2\sqrt{3}}$$

It's a good idea to rationalise the denominator by multiplying the top and bottom of the fraction by $\sqrt{3}$.

Since the ladder is on the point of slipping, friction is limiting and the coefficient of friction is given by:

$$\mu = \frac{F}{\mathbf{R}_A} = \frac{7}{10\sqrt{3}} = \frac{7\sqrt{3}}{30}$$

Progress check

1 A uniform plank AB of length 3 m and mass 25 kg rests horizontally on two supports, one at each end. A mass of 15 kg is positioned on the plank 1 m from B. Find the reaction at each of the supports. Take $g = 9.8$ ms^{-2}.

2 A uniform ladder of length 5 m rests against a smooth wall with its base on rough horizontal ground. The coefficient of friction between the ladder and the ground is $\frac{2}{3}$ and the ladder is on the point of slipping. Show that the angle between the ladder and the ground is $\tan^{-1}(\frac{3}{4})$.

1   220.5 N (A), 171.5 N(B)

3.3 Centre of mass

After studying this section you should be able to:

- find the centre of mass of a system of particles in one and two dimensions
- find the centre of mass of a lamina by considering an equivalent system of particles
- use centre of mass to determine conditions for the equilibrium of a lamina

Centre of mass of a system of particles

In the diagram, m_1 and m_2 lie on a straight line through O. Their displacements from O are x_1 and x_2 respectively.

The position given by $\bar{x} = \dfrac{m_1 x_1 + m_2 x_2}{m_1 + m_2}$ is called the **centre of mass** of m_1 and m_2.

> If m_1 and m_2 are equal then this result gives the mid-point of the two masses.

For n separate masses m_1, m_2, ... m_n the position of the centre of mass is given by

$$\bar{x} = \frac{m_1 x_1 + m_2 x_2 + \ldots m_n x_n}{m_1 + m_2 + \ldots m_n}$$ This is usually written as $\bar{x} = \dfrac{\displaystyle\sum_{i=1}^{n} m_i x_i}{\displaystyle\sum_{i=1}^{n} m_i}$

In two dimensions, using the usual Cartesian coordinates, the position of the centre of mass is at $(\bar{x}, \bar{y})$ where $\bar{x}$ and $\bar{y}$ are given by

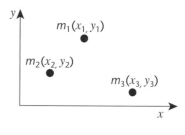

$$\bar{x} = \frac{\displaystyle\sum_{i=1}^{n} m_i x_i}{\displaystyle\sum_{i=1}^{n} m_i} \text{ as above, and } \bar{y} = \frac{\displaystyle\sum_{i=1}^{n} m_i y_i}{\displaystyle\sum_{i=1}^{n} m_i}$$

Example

Masses of 2 kg, 3 kg and 5 kg are positioned as shown in the diagram.
Find the coordinates of the centre of mass of the system.

> The method is easily extended to deal with any number of masses.

$$\bar{x} = \frac{2 \times 2 + 3 \times 4 + 5 \times 7}{2 + 3 + 5} = \frac{51}{10} = 5.1$$

$$\bar{y} = \frac{2 \times 1 + 3 \times 5 + 5 \times 3}{2 + 3 + 5} = \frac{32}{10} = 3.2$$

The centre of mass has coordinates (5.1, 3.2)

Centre of mass of a lamina

A **lamina** is something that is thin and flat such as a sheet of metal. The thickness of a lamina is ignored and it is treated as a two-dimensional object.

You can find the centre of mass of a uniform lamina by dividing it into parts which you then represent as particles. Each part is usually rectangular or triangular and in order to represent it as a particle:

- the mass of the particle is represented by the area of the part
- the position of the particle is taken to be at the geometric centre of the part.

This representation of a lamina gives you the information you need to apply the formulae for its centre of mass.

Example

Find the coordinates of the centre of mass of this uniform lamina.

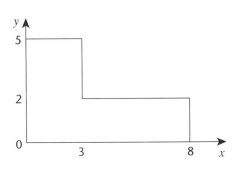

There are different ways to do this, but it doesn't matter which you choose.

The first step is to divide the lamina into rectangles.

This process is easily extended to deal with more complex composite figures.

Rectangle A has area 15 square units. Its centre is at (1.5, 2.5).

Rectangle B has area 10 square units. Its centre is at (5.5, 1).

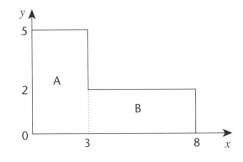

$$\bar{x} = \frac{15 \times 1.5 + 10 \times 5.5}{15 + 10} = 3.1$$

$$\bar{y} = \frac{15 \times 2.5 + 10 \times 1}{15 + 10} = 1.9$$

The centre of mass of the lamina has coordinates (3.1, 1.9).

Freely suspended lamina

Once you know where the centre of mass of a lamina is, you can use this to work out how it will move if it is freely suspended from a given point.

> When a lamina is freely suspended, it will move so that its centre of mass lies directly below the point of suspension.
>
> **KEY POINT**

Example

The diagram shows the lamina of the previous example hanging freely from one corner.

Find the size of angle θ between the direction of Oy and the horizontal.

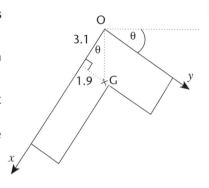

The position of the centre of mass is shown at the point G.
Notice that θ also lies in a right-angled triangle in which two of the sides are known.

From the diagram: $\tan \theta = \dfrac{1.9}{3.1}$

giving $\theta = 31.5°$ to 1 d.p.

Progress check

1 Masses of 2 kg, 3 kg and 5 kg are positioned as shown in the diagram.

   (a) Find the coordinates of G, the centre of mass of the system.
   (b) Find the angle θ between OG and Ox.

3.4 Collisions and impulse

After studying this section you should be able to:

- *apply the principle of conservation of momentum to direct impact*
- *apply Newton's experimental law using the coefficient of restitution*
- *calculate the loss of mechanical energy in a collision*
- *recall and use the definition of impulse as a change of momentum*

LEARNING SUMMARY

Conservation of momentum

The **momentum** of an object is the product of its mass and its velocity, i.e. momentum = $m\mathbf{v}$. Momentum has direction and so it is a vector quantity.

Momentum is measured in Ns (Newton seconds).

> It's important to use the correct units. Remember that speed is the magnitude of the velocity.

For example, an object of mass 5 kg moving with speed 6 ms^{-1} in a given direction has momentum $5 \times 6 = 30$ Ns in the direction of movement.

The **principle of conservation of momentum** states that, in a collision, the total momentum before impact = the total momentum after impact.

Key points from AS

- **Momentum**
  Revise AS page 67

Remember:

- always draw a diagram to represent the situation
- choose one direction to be positive (usually left to right $\rightarrow$)
- show any unknown velocities in the positive direction.

Example

> When particles coalesce, they stick together and behave as a single particle.

Two particles A and B of masses 4 kg and 6 kg respectively move towards each other along the same straight line. Particle A is moving with speed 3 ms^{-1} and particle B is moving with speed 5 ms^{-1}. Given that the particles coalesce on impact, find their common velocity immediately afterwards.

> The velocity of B is in the negative direction.

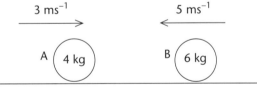

Before impact:

After impact: v ms^{-1}

By the principle of conservation of momentum:

$$4 \times 3 - 6 \times 5 = 10v$$

giving: $10v = -18$

so: $v = -1.8$

> v is negative so the direction of the velocity is from right to left.

The particles move with velocity 1.8 ms^{-1} in the direction from B to A.

Newton's experimental law

Some problems require two unknown values to be found and in this situation a second equation is needed.

Newton found by experiment that, when two objects collide, the ratio of their speed of separation to their speed of approach has a fixed value. The symbol used to represent this ratio is e and its value is depends on what the objects are made of.

$$e = \frac{\text{speed of separation}}{\text{speed of approach}}$$

This is known as **Newton's experimental law** and is often used to provide a second equation.

This is often used in the form $e \times$ speed of approach = speed of separation.

The speed of separation is always less than or equal to the speed of approach so it follows that $0 \leqslant e \leqslant 1$. The ratio e is called the **coefficient of restitution**.

Example

In the exam, you may need to deal with more than one impact. See the sample exam questions page 102.

A smooth sphere P moves towards a stationary smooth sphere Q in a straight line through their line of centres. P has mass 2 kg and speed 5 ms^{-1}. Q has mass 4 kg. Given that the coefficient of restitution between the spheres is 0.5, find:

(a) the speed of each sphere immediately after impact
(b) the loss of energy in the collision.

(a)

Before impact:

5 ms^{-1}

After impact:

P (2 kg)   Q (4 kg)

Momentum is conserved:
$2 \times 5 + 4 \times 0 = 2v_1 + 4v_2$

v_1 v_2

So: $\qquad\qquad\qquad v_1 + 2v_2 = 5 \qquad\qquad (1)$

Newton's law: $\qquad\qquad 0.5 \times 5 = v_2 - v_1 \qquad\qquad (2)$

Adding (1) and (2) gives: $\qquad 3v_2 = 7.5 \implies v_2 = 2.5$
and: $\qquad\qquad\qquad\qquad\qquad\qquad v_1 = 0$

Immediately after the impact, P is brought to rest and Q moves with speed 2.5 ms^{-1}.

Kinetic energy $= \frac{1}{2}mv^2$. Even though momentum is conserved, energy is lost.

(b) Loss of energy = KE before impact − KE after impact
$$= 0.5 \times 2 \times 5^2 - 0.5 \times 4 \times 2.5^2$$
$$= 12.5 \text{ J}$$

Impulse

The **impulse** of a constant force acting over a given time is given by the product of force and time. This may be written as $I = Ft$ where the impulse is I N s, the force is F N and the time is t s.

It follows that: Impulse = change in momentum

so $Ft = mv - mu$.

Example

A particle of mass 4 kg, initially at rest, is acted upon by a force of 3 N for 10 s. What is the speed of the particle at the end of this time?

Using $Ft = mv - mu$

gives $3 \times 10 = 4v - 0$

$\Rightarrow$ $v = 7.5$

The speed of the particle is 7.5 m s^{-1}.

When two particles collide, the contact force between them may only last a very short time and is unlikely to be constant. However, the value of F in the equation $Ft = mv - mu$ may be used to represent the average value of this force.

Example

A ball of mass 0.5 kg strikes a hard floor with speed 2 m s^{-1} and rebounds with speed 1.5 m s^{-1}.
Given that the ball is in contact with the floor for 0.05 s find the average value of the contact force.

Taking upwards as the positive direction:

$u = -2$, $v = 1.5$, $m = 0.5$ and $t = 0.05$

Using $Ft = mv - mu$

gives $0.05F = 0.5 \times 1.5 - 0.5 \times (-2)$

$\Rightarrow$ $0.05F = 1.75$

$\Rightarrow$ $F = 35$

The average value of the contact force is 35 N.

Impulse = change in momentum

When two particles collide, each receives an impulse from the other of equal size but opposite sign. In this way, the total change in momentum is zero. We would expect this to happen because the total momentum of the system is conserved in a collision.

The total momentum of the system is conserved.

Example

A ball of mass 0.5 kg moving with velocity $6\mathbf{i} + 2\mathbf{j}$ ms^{-1} is kicked and receives an impulse of $2\mathbf{i} + 11\mathbf{j}$ Ns. Find the velocity of the ball immediately after it is kicked.

Momentum immediately after kick = initial momentum + impulse

$$= 0.5(6\mathbf{i} + 2\mathbf{j}) + 2\mathbf{i} + 11\mathbf{j} = 5\mathbf{i} + 12\mathbf{j}$$
$$= 0.5\mathbf{v} \text{ (where } \mathbf{v} \text{ is the velocity)}$$

This gives:
$$\mathbf{v} = 10\mathbf{i} + 24\mathbf{j} \text{ ms}^{-1}$$

Progress check

1 Two particles A and B move towards each other with speeds of 6 ms^{-1} and 2 ms^{-1} respectively. A has mass 3 kg and B has mass 1 kg. The coefficient of restitution between the particles is 0.6. Find the speed of each particle after the collision.

2 A particle of mass 8 kg moving with speed 2 m s^{-1} hits a stationary particle of mass 2 kg.
   After the impact both particles move in the same direction with speed v m s^{-1}.
   (a) Find the value of v.
   (b) Find the impulse given to the stationary particle.

2 (a) $v = 1.6$ (b) Impulse = 3.2 N s.
1 A 2.8 ms^{-1}, B 7.6 ms^{-1}.

3.5 Uniform circular motion

After studying this section you should be able to:

- *understand angular speed and be able to apply the formula $v = r\omega$*
- *understand that a particle moving in a circular path with constant speed has an acceleration towards the centre of the circle*
- *use the formula $\omega^2 r$, or its equivalent form $\dfrac{v^2}{r}$, to represent the magnitude of the acceleration towards the centre of the circle in solving problems*

LEARNING SUMMARY

Angular speed

Suppose that a particle P moves with constant speed in a circular path with centre O.

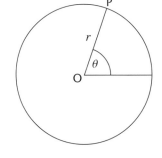

The direction of OP changes with time.
The angle θ, through which OP turns in some given time, is usually measured in radians.

The **angular speed** of P is the rate of change of θ with respect to time. Using ω to represent angular speed in radians per second gives:

$$\omega = \frac{\theta}{t}$$

One advantage of using radians is that the relationship between angular speed ω, and linear speed v, is easily expressed as $v = r\omega$.

For example, a particle moving in a circular path of radius 2 m with angular speed 10 radians/sec has a straight line speed of $2 \times 10 = 20$ ms^{-1}.

Radial acceleration

Notice that while the speed of the particle may be constant, its *direction* changes as it moves around the circle. This means that the *velocity* is *not constant* and so the particle must be *accelerating*.

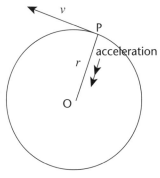

This acceleration is always directed towards the centre of the circle and is known as **radial acceleration** or sometimes **centripetal acceleration**.

The magnitude of this acceleration is given by $a = \omega^2 r$.

Since $v = r\omega$, it follows that $a = \dfrac{v^2}{r}$.

This means that there is a choice of which formula to use. In practice, you should use the one that best fits the available information.

Example

Since we are given the angular speed, the formula $a = \omega^2 r$ is the simpler one to use here.

A particle moves in a circular path of radius 3 m with angular speed 5 radians/sec. Find the magnitude of its radial acceleration.

Using $a = \omega^2 r$:   the radial acceleration is $5^2 \times 3 = 75$ ms^{-2}.

Circular motion and force

The radial acceleration of an object moving in a circular path must be produced as the result of a *force acting towards the centre of the circle*. This force is sometimes called the **centripetal force**.

Using $F = ma$, the force needed to keep an object of mass m kg moving in a circular path is $m\omega^2 r$. An alternative form of this is $\dfrac{mv^2}{r}$.

In applying this theory to solving problems, the centripetal force may take various forms such as a tension in a string, friction or the gravitational attraction of a planet.

Example

One end of a string of length 80 cm is attached to a point O on a smooth horizontal table. The other end is attached to a mass of 3 kg moving with speed 4 ms^{-1} and the string is taut. Find the tension in the string.

> It is important to write the length of the string in metres.

Using $T = \dfrac{mv^2}{r}$ gives $T = \dfrac{3 \times 4^2}{0.8} = 60$

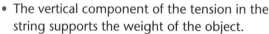

The tension in the string is 60 N.

The conical pendulum

One particular situation involving circular motion is described as the **conical pendulum**.

This is where an object moves in a horizontal circle while suspended by a string attached to a point vertically above the centre of the circle.

- The vertical component of the tension in the string supports the weight of the object.
- The horizontal component of the tension in the string provides the centripetal force.

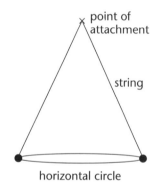

Example

A light inextensible string is fixed at one end to a point O. The other end is attached to a particle of mass 2 kg which moves in a horizontal circle of radius 30 cm with angular speed 4 radians per second. Find the angle of inclination of the string to the vertical. Take $g = 9.8$ ms^{-2}.

Vertically: $T \cos \theta = 19.6$ (1)

Horizontally: $T \sin \theta = 2 \times 4^2 \times 0.3 = 9.6$ (2)

(2) ÷ (1) gives: $\tan \theta = \dfrac{9.6}{19.6} \Rightarrow \theta = 26.1°$

The string is inclined at 26.1° to the vertical.

Progress check

1 One end of a string of length 60 cm is attached to a point O on a smooth horizontal table. The other end is attached to a mass of 5 kg moving with speed 3 ms^{-1} and the string is taut. Find the tension in the string.

3.6 Work, energy and power

After studying this section you should be able to:

- *understand kinetic and potential energy and the work energy principle*
- *understand the principle of conservation of mechanical energy*
- *understand the definition of power*
- *solve problems involving work, energy and power*

LEARNING SUMMARY

Work and energy

In everyday language, energy is needed to do work. This idea is given a formal interpretation within mechanics.

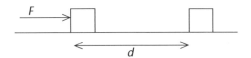

The work done, W, by a *constant* force, F, in moving an object through a distance, d, along its line of action is given by $W = Fd$. The unit of work is the joule (J).

Energy is also measured in joules and the energy used in moving the object is equal to the work done. In a way, work and energy are like two sides of the same coin.

Energy may take different forms such as heat, sound, electrical, chemical and mechanical. An important property of energy is that it may change from one form to another. For your mechanics module, the focus of attention is on mechanical energy.

There are two forms of mechanical energy: kinetic energy (KE)
and potential energy (PE)

The kinetic energy of an object is the energy it has due to its *motion*.

Its value depends on the mass of the object and its speed. $\text{KE} = \frac{1}{2}mv^2$.

For example, a mass of 4 kg moving with speed 5 ms^{-1} has $\text{KE} = \frac{1}{2} \times 4 \times 5^2 = 50$ J

The potential energy of an object is the energy it has due to its *position*.

The gravitational potential energy of an object, relative to some level, is equal to the work done against gravity in raising the object from that level to its current position.

PE = *mgh*
relative to ground level.

The total mechanical energy of an object is the sum of its KE and PE.

The **principle of conservation of mechanical energy** states that the total mechanical energy of a system remains constant, provided that it is not acted on by any external force, other than gravity.

KEY POINT

The principle of conservation of energy applies, for example, to the motion of a projectile. Any change in the PE of the projectile corresponds to an equal but opposite change in its KE.

Power

Power is the rate of doing work. It is measured in watts (W) where $1\ W = 1\ Js^{-1}$.

If the point of application of a force F moves with speed v in the *direction of the force* then the power P is given by $P = Fv$.

Example

Find the power generated by a car engine when the car travels at a constant speed of 72 kmh^{-1} on horizontal ground against resistance forces of 2500 N.

> Since the speed is constant, the force produced by the engine must match the resistance force.

$$72\ kmh^{-1} = \frac{72 \times 1000}{60 \times 60}\ ms^{-1} = 20\ ms^{-1}.$$

$$\text{Power} = 2500 \times 20\ W = 50\ 000\ W$$
$$= 50\ kW$$

Exam questions may involve resisted motion on an inclined plane.

Example

A car of mass 1000 kg travels up a hill inclined at angle θ to the horizontal, where $\sin \theta = \frac{1}{20}$. The non-gravitational resistance to motion is 2000 N and the power output from the engine is 60 kW. Find the acceleration of the car when it is travelling at 10 ms^{-1}. Take $g = 10\ ms^{-2}$.

> A clearly labelled diagram is a must.

> In this formula, F represents the *resultant force* up the hill.

Using $P = Fv$: $60\ 000 = F \times 10 \Rightarrow F = 6000$

Using $F = ma$: $6000 - 2000 - 10\ 000 \times \frac{1}{20} = 1000a$

This gives: $3500 = 1000a \Rightarrow a = 3.5$

The acceleration of the car up the hill is 3.5 ms^{-2}.

Progress check

1 A ball is kicked from ground level with speed 15 ms^{-1} and just clears a wall of height 2 m. The speed of the ball as it passes over the wall is v ms^{-1}.

   (a) Assume that the ball has mass m kg and use the conservation of mechanical energy principle to write an equation.

   (b) Solve the equation to find v.

2 A car of mass 800 kg travels up a hill inclined at angle θ to the horizontal, where $\sin \theta = \frac{1}{20}$. The non-gravitational resistance to motion is 1600 N and the power output from the engine is 45 kW. Find the acceleration of the car when it is travelling at 15 ms^{-1}. Take $g = 10\ ms^{-2}$.

2 1.25 ms^{-2}
1 (b) 13.6 ms^{-1}
   (a) $\frac{1}{2}mv^2 = m \times 9.8 \times 2 + \frac{1}{2}m \times 15^2$

Sample questions and model answers

1

A ball of mass 250 g is released from rest 2 m above the ground. The coefficient of restitution between the ball and the ground is 0.7. Take $g = 9.8 \text{ ms}^{-2}$.

(a) Find the height reached by the ball as it rebounds from the ground.

(b) Find the impulse exerted on the ball by the ground.

(a) Using $v^2 = u^2 + 2as$

gives $v^2 = 2 \times 9.8 \times 2$

$= 39.2$

so $v = \sqrt{39.2}$

There is no need to work out $\sqrt{39.2}$ at this stage.

The ball strikes the floor with speed $\sqrt{39.2} \text{ ms}^{-1}$.

Using Newton's law, the ball rebounds with speed $\sqrt{39.2} \times 0.7 \text{ ms}^{-1}$

Some simple statements help to make your method clear.

Using $v^2 = u^2 + 2as$

gives $0 = 39.2 \times 0.49 - 2 \times 9.8 \times h$

so $h = \dfrac{39.2 \times 0.49}{2 \times 9.8} = 0.98$

The ball reaches a height of 0.98 m when it rebounds from the ground.

(b) Taking upwards as positive:

Momentum of ball immediately before it hits the ground
$= -0.25 \times \sqrt{39.2} \text{ Ns}$

Write the mass in kg.

Momentum of ball immediately after it hits the ground
$= 0.25 \times \sqrt{39.2} \times 0.7 \text{ Ns}$

$-0.25 \times \sqrt{39.2} \text{ Ns} \qquad 0.25 \times \sqrt{39.2} \times 0.7 \text{ Ns}$

Impulse $=$ change in momentum

$= 0.25 \times \sqrt{39.2} \times 0.7 - (-0.25 \times \sqrt{39.2})$

$= 2.66 \text{ Ns to 3 s.f.}$

The impulse exerted on the ball by the ground is 2.66 Ns.

Sample questions and model answers (continued)

2

A particle P of mass $3m$ is moving with speed $2u$ when it strikes a particle Q of mass m which is at rest. The coefficient of restitution between the particles is e.

(a) Show that the speed of P after the collision is $\frac{v}{2}(3 - e)$ and find the speed of Q.

(b) Q subsequently strikes a wall perpendicular to its direction of motion and rebounds so that a second collision with P occurs. Given that the coefficient of restitution between Q and the wall is $\frac{1}{3}$ find the speed of P after the second collision.

(c) Show that P moves in the same direction after the second collision.

(a) Before impact

$$\overset{2u}{\longrightarrow}$$

P $\left(\, 3m \,\right)$ Q $\left(\, m \,\right)$

After impact

$$\overset{}{\underset{v_P}{\longrightarrow}} \qquad \overset{}{\underset{v_Q}{\longrightarrow}}$$

> The approach is exactly the same as when numerical values of the speeds are given.

Conservation of momentum gives: $6mu = 3mv_P + mv_Q$

which simplifies to $\qquad\qquad 6u = 3v_P + v_Q$ (1)

Newton's experimental law gives: $2ue = v_Q - v_P$ (2)

(1) − (2) gives: $\qquad\qquad 6u - 2ue = 4v_P$

> This is the result given in the question.

giving $\qquad\qquad\qquad v_P = \frac{u}{2}(3 - e)$ as required

and from (2) $\qquad v_Q = 2ue + v_P = 2ue + \frac{u}{2}(3 - e)$

$$= \frac{u}{2}(4e + 3 - e) = \frac{3u}{2}(e + 1)$$

The speed of Q after the collision is $\frac{3u}{2}(e + 1)$.

Sample questions and model answers (continued)

(b) The speed of Q after hitting the wall is $\dfrac{1}{3} \times \dfrac{3u}{2}(e+1) = \dfrac{u}{2}(e+1)$ in the opposite direction.

Repeat the process with the new speeds. Draw a diagram and set your working out clearly.

$$\dfrac{u}{2}(3-e) \longrightarrow \qquad\qquad \dfrac{u}{2}(e+1) \longleftarrow$$

Before impact   P ⟨3m⟩   Q ⟨m⟩

After impact $\longrightarrow w_P$ $\longrightarrow w_Q$

Conservation of momentum gives: $\dfrac{3mu}{2}(3-e) - \dfrac{mu}{2}(e+1) = 3mw_P + mw_Q$

which simplifies to $\qquad\qquad\qquad\qquad 4u - 2ue = 3w_P + w_Q \qquad (3)$

Newton's experimental law gives: $\left(\dfrac{u}{2}(3-e) + \dfrac{u}{2}(e+1) \right)e = w_Q - w_P$

which simplifies to $\qquad\qquad\qquad\qquad\qquad 2u = w_Q - w_P \qquad (4)$

(3) − (4) gives $\qquad\qquad\qquad\qquad\qquad\qquad 2u - 2ue = 4w_P$

and $\qquad\qquad\qquad\qquad\qquad\qquad\qquad w_P = \dfrac{u}{2}(1-e)$

There is no need to find w_Q.

so the speed of P immediately after the second collision is $w_P = \dfrac{u}{2}(1-e)$.

(c) $1-e \geqslant 0$ since $0 \leqslant e \leqslant 1$ so the direction of movement of P is not changed.

Sample questions and model answers (continued)

3

The diagram shows a particle of mass 0.2 kg threaded onto a smooth string fixed at its end-points A and B. The points A, B and C lie in a vertical line and the particle travels in a horizontal circle of radius 0.5 m with centre C.

(a) Calculate the tension in the string.

(b) Calculate the speed of the particle.

(a) Since the string is smooth, the tension in both parts of the string is the same.

The diagram shows the forces acting on the particle.

There is no acceleration vertically.

Resolving vertically $\qquad T\cos 40° + T\cos 80° = 0.2 \times 9.8$

$$T = \frac{0.2 \times 9.8}{0.93969} = 2.0857\ldots$$

The tension in the string is 2.09 N (3 s.f.)

(b) Let the speed of the particle be v ms^{-1}.

Resolving horizontally $\qquad$ Centripetal force $= T\sin 40° + T\sin 80°$

Centripetal force $= \dfrac{mv^2}{r}$

$$= 3.3948\ldots = \frac{0.2v^2}{0.5}$$

$$v = \sqrt{\frac{3.3948}{0.4}}$$

$$= 2.913\ldots$$

The speed of the particle is 2.91 ms^{-1} (3 s.f.)

Practice examination questions

1 A ball is struck with speed 25 ms^{-1} from a point 50 cm above the ground. When it has travelled 20 m horizontally, the ball just clears a fence that is 3 m high.

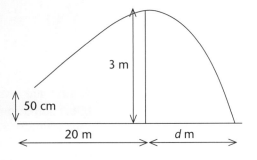

(a) Show that a possible angle of projection of the ball is 16.5°.

(b) Find the distance d m, between the fence and the point where the ball strikes the ground for this angle of projection.

2 The diagram shows a uniform lamina ABCDEFGH.

(a) Write down the distance of the centre of mass of the lamina from AH.

(b) Find the distance of the centre of mass of the lamina from AB.

(c) The lamina is now suspended freely from A. Find the angle that AB makes with the horizontal in its equilibrium position.

3

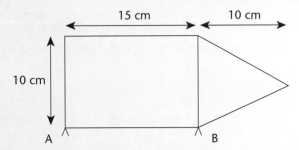

The diagram shows a uniform lamina of mass 0.8 kg resting on supports at A and B.

(a) Find the horizontal distance of the centre of mass from the support at A.

(b) Find the magnitude of the reaction force on the lamina at each support.

4

A and B are two smooth spheres of equal size and they are approaching each other on a direct line through their centres. A is moving with speed 4 ms⁻¹ and B is moving with speed 2 ms⁻¹. The coefficient of restitution between the spheres is $\frac{1}{3}$.

Find:

(a) the speed of each sphere after the collision

(b) the impulse exerted by A on B during the collision

(c) the loss of mechanical energy during the collision.

5 Two particles P and Q approach each other with speeds u and $2u$ respectively. P has mass $3m$ and Q has mass m. The coefficient of restitution between the particles is e.

Given that the direction of motion of P is reversed in the collision, show that $e > \frac{1}{3}$.

6

A particle P of mass $2m$ moving with speed $5u$ strikes particle Q of mass m, which is at rest. After the collision both particles move in the same direction but the speed of particle Q is twice the speed of particle P.

(a) Find the speed of particle P after the collision.

(b) Find the magnitude of the impulse exerted on Q by P during the impact.

7 A motorbike is driven round a corner in a circular path of radius 40 m. The maximum speed that this can be done without skidding is 15 ms^{-1}.

Calculate the coefficient of friction between the tyres and the road surface to 2 s.f.

Take $g = 9.8$ ms^{-2}.

8 One end of a light inelastic string of length 50 cm is attached to a fixed point 40 cm above a smooth horizontal table. The other end is attached to an object of mass 2 kg moving in a circular path of radius 30 cm on the table.

Given that the angular speed of the object is 3 radians per second, find:

(a) the magnitude of the acceleration of the object

(b) the tension in the string

(c) the reaction force of the table on the object.

9 A particle P of mass m is attached to one end of a light inelastic string of length l. The other end of the string is attached to a fixed point O. P is released from rest with the string taut and horizontal.

(a) Explain why, in the subsequent motion, the tension in the string does no work on the particle.

(b) Explain why it is not valid to use the formula $v^2 = u^2 + 2as$ in this situation.

(c) Find the maximum speed of the particle assuming that there is no resistance to motion.

10 A car of mass 900 kg travels up a straight road inclined at angle α to the horizontal, where $\sin \alpha = \frac{1}{15}$. Assume that the total non-gravitational force opposing the car's motion has a constant value of 800 N. Take $g = 9.8$ ms^{-2}.

(a) Find the driving force required for the car to maintain a steady speed up the slope.

(b) Find the power output of the car's engine given that its speed up the slope has a constant value of 20 ms^{-1}.

(c) At the top of the hill, the road surface is horizontal. If the power output of the engine remains the same, what is the initial acceleration of the car?

11 A football of mass 0.2 kg is travelling horizontally towards the goal with speed 20 ms^{-1} when it is struck by the keeper. It then rebounds vertically with speed 5 ms^{-1}.

Find the magnitude and direction of the impulse given to the ball by the keeper.

Practice examination questions *(continued)*

12 A car is driven round a track in a horizontal circle of radius 80 m at a speed of 20 ms⁻¹. The track is banked at angle α to the horizontal and there is no tendency for the car to side-slip.

By representing the car as a particle:

(a) Find α to 1 d.p.

(b) The maximum speed that the car can travel on the banked track without slipping sideways is 28 ms⁻¹. Find the coefficient of friction between the tyres and the surface of the track.

13 The diagram shows a particle of mass 0.2 kg attached to one end of a light inelastic string of length 80 cm.

The particle moves in a horizontal circle on a smooth surface with the string inclined at 30° to the vertical.

(a) Given that the speed of the particle is 1.2 ms⁻¹ find the tension in the string and the reaction of the table on the particle.

(b) Find the minimum angular speed of the particle so that the reaction of the table on the particle is zero.

Chapter 4

Decision Mathematics 2

The following topics are covered in this chapter:

- Game theory
- Networks
- Critical path analysis

- Matching and allocation
- Dynamic programming

4.1 Game theory

After studying this section you should be able to:

- understand the idea of a zero-sum game and its representation using a pay-off matrix
- identify play-safe strategies and stable solutions
- find optimal mixed strategies for a game with no stable solution
- reduce a pay-off matrix using a dominance argument

LEARNING SUMMARY

A two person zero-sum game

The starting point in the mathematical theory of games is that the outcome of a game is determined by the **strategies** of the players.

A two person **zero-sum game** is a game in which the winnings of one player equal the losses of the other for every combination of strategies. Taking winnings to be positive and losses to be negative gives a zero sum in each case.

Viewing a game from one player's point of view, we could represent the outcomes (called pay-offs) for each combination of strategies in a matrix. This is called the **pay-off matrix** for that player.

Example

A and B are two players in a zero-sum game. A uses one of two strategies, W or X, and B uses one of the strategies Y or Z. The table shows the pay-off matrix for A.

The situation could equally be represented by the pay-off matrix for B. This would show corresponding values with opposite signs since this is a zero-sum game.

Pay-off matrix for A		B	
		Y	Z
A	W	2	−2
	X	5	−4

The pay-off matrix shows that if B adopts strategy Y then the pay-off for A will be 2 by using strategy W and 5 using strategy X. On the other hand, if B adopts strategy Z then the pay-off for A will be −2 using strategy W and −4 using strategy X.

The idea is that neither player knows in advance which strategy the other will use.

The play-safe strategy

The **play-safe** strategy for a player is the strategy for which the minimum pay-off is as high as possible. In the example above, the minimum pay-off for A using strategy W is −2, whereas the minimum pay-off using strategy X is −4. This means that strategy W is the play-safe strategy for player A.

Notice that finding the play-safe strategy for player A involves comparing the minimum values in the *rows* of the pay-off matrix for A. Finding the play-safe strategy for B will involve comparing the values in the *columns*, remembering that B's pay-offs are the negatives of the ones in the pay-off matrix for A.

The minimum value for B using strategy Y is −5 and the minimum value using strategy Z is 2. This means that the play-safe strategy for B is strategy Z.

The situation is shown in the pay-off matrices for A and B.

Pay-off matrix for A		B Y	Z
A	W	2	−2
	X	5	−4

Pay-off matrix for B		B Y	Z
A	W	−2	2
	X	−5	4

In this case, the maximum of the minimum pay-offs, for each player, is in the corresponding position in the two matrices. This represents the **stable solution** to the problem referred to as the **saddle point** (or **minimax point**).

The solution is stable in the sense that neither player can improve their pay-off by taking a different strategy, given that the other player doesn't change. In other words, while B uses strategy Z, the best strategy for A is W and while A uses strategy W, the best strategy for B is Z.

> In the example, the sum of these values is −2 + 2 = 0 and the solution is stable.

> **KEY POINT**
>
> If the sum of the two values used to determine the play-safe strategies is **not zero** then the values cannot correspond to the same cell position in the play-off matrices. This means that there is no saddle point and the game has **no stable solution**.

Example

The pay-off matrices for two players in a zero-sum game are given below. Show that there is no stable solution for the game.

Pay-off matrix for P		Q Y	Z
P	W	5	−3
	X	−4	6

Pay-off matrix for Q		Q Y	Z
P	W	−5	3
	X	4	−6

The play-safe strategies for P and Q are shown shaded.
The values used to determine the play-safe strategies are −3 and −5.
There is no stable solution since $(-3) + (-5) \neq 0$.

Optimal strategies for games that are not stable

The repeated use of the same strategy over a series of games is called a **pure strategy**. This provides the best results for both players in a game which has a stable solution. In the case where no stable solution exists, a **mixed strategy** is used in which each of the strategies is employed with a given probability to find the optimal solution.

Returning to the previous example:

Pay-off matrix for P	Q	
	Y	Z
P W	5	−3
X	−4	6

Suppose that P chooses:

      strategy W with probability p

and   strategy X with probability $(1 - p)$.

Then if:

Q chooses strategy Y, the expected *gain* for P is given by $5p - 4(1 - p) = 9p - 4$

Q chooses strategy Z, the expected *gain* for P is $-3p + 6(1 - p) = -9p + 6$

> The optimal value occurs when these expressions are equal.

$$9p - 4 = -9p + 6 \Rightarrow 18p = 10$$
$$\Rightarrow p = \tfrac{5}{9}$$

> You could equally use $-9p + 6$ to get the value of the game.

The value of the game is given by $9p - 4 = 1$.

Suppose that Q chooses:

      strategy Y with probability q

and   strategy Z with probability $(1 - q)$.

Then if:

P chooses strategy W, the expected *loss* for Q is given by $5q - 3(1 - q) = 8q - 3$

P chooses strategy X, the expected *loss* for Q is $-4q + 6(1 - q) = -10q + 6$

$$8q - 3 = -10q + 6 \Rightarrow 18q = 9$$
$$\Rightarrow q = \tfrac{1}{2}$$

(As a check, the value of the game is $8q - 3 = 1$ as before).

This shows that the optimal strategy for both players is to use mixed strategies such that:

P chooses strategy W with probability $\tfrac{5}{9}$ and strategy X with probability $\tfrac{4}{9}$

Q chooses strategy Y and strategy Z with equal likelihood.

The expected long-term gain then for P, as an average per game, is 1 and this is also the expected long-term loss, as an average per game, for Q.

Graphical representation

Each graph corresponds to a strategy for Q (i.e. the opponent of P).

The diagram shows graphs of $9p - 4$ and $-9p + 6$ against values of p from 0 to 1.

The point of intersection corresponds to the probability that gives the optimal mixed strategy for P in the last example.

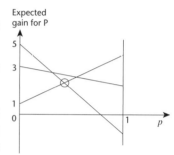

> **KEY POINT**
>
> When P's opponent has more strategies there will be more graphs with several points of intersection. You will need to identify the one that represents the optimal mixed strategy for P.
> *This will be the highest point on or below each of the graphs.*

This diagram represents a situation where P's opponent has three strategies to choose from.

The point representing the optimal mixed strategy for P is circled.

Notice how the problem of identifying the right vertex can be expressed as a linear programming problem in which the object is to maximise the expected gain for P subject to the constraints represented by the regions bounded by the straight line graphs.

This is, in fact, the approach used for higher dimensional problems. The conditions are formulated as a linear programming problem which may then be solved by the Simplex algorithm.

> **KEY POINT**
>
> If the pay-offs for one strategy are *always* better than the corresponding pay-offs for some other strategy then the weaker one can be ignored when determining the probabilities for a mixed strategy. In this way, the pay-off matrix is reduced by what is called a **dominance argument**.

Progress check

1 The table shows a pay-off matrix for player A in a zero-sum game.

(a) Find the play-safe strategy for each player.

(b) Show that this game has no stable solution.

(c) Find the best strategy for each player and the value of the game.

Pay-off matrix for A	Q	
	Y	Z
A W	6	−4
A X	−3	5

1 (a) A(strategy X), Q(strategy Z)
(b) $(-3) + (-5) \neq 0$
(c) A use W and X with probabilities $\frac{4}{9}$ and $\frac{5}{9}$
Q use Y and Z with equal probability, value of game = 1

4.2 Networks

After studying this section you should be able to:

- *understand flows in networks and be able to find the maximum flow for a network involving multiple sources and sinks*

Flows in networks

The **flows** referred to may be flows of liquids, gases or any other measurable quantities. The edges may represent such things as pipes, wires or roads that carry the quantities between the points identified as vertices.

A typical vertex has a flow into it and a flow out of it. The exceptions are a **source** vertex which has no input and a **sink** vertex which has no output.

Each edge of the network has a **capacity** which represents the maximum possible flow along that edge. In the usual notation, the capacity is written next to the edge and the flow is shown in a circle.

This edge has a capacity of 10 and a flow of 7.

The set of flows for a network is **feasible** if:

- The total output from all source vertices is equal to the total input for all sink vertices.
- The input for each vertex other than a source or sink vertex is equal to its output.
- The flow along each edge is less than or equal to its capacity.

A **cut** divides the vertices into two sets, one set containing the source and the other containing the sink.

The **capacity of a cut** is equal to the sum of the capacities of the edges that cross the cut, *taken in the direction from the source set to the sink set*.
In the diagram, the capacity of cut (i) is 15 + 10 = 25.

> Be careful to add the capacities and not the flows.

The capacity of cut (ii) is 8 + 14 = 22.
Notice that, for the second cut, the capacity of 4 is not included because it does not go from the source set to the sink set.

The maximum flow–minimum cut theorem

The **maximum flow–minimum cut theorem** states that the maximum value of the flow through a network is equal to the capacity of the minimum cut. (This is a little bit like saying that a chain is only as strong as its weakest link.)

It follows from the theorem that if you find a flow that is equal to the capacity of some cut then it must be the maximum flow that can be established through the network.

For example, the capacity of cut (ii) for this network has already been found to be 22. The circled figures show one way of achieving this flow and the theorem now tells us that this must be the maximum possible flow through the network.

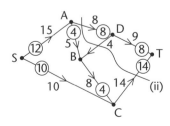

Flow augmentation

Once an initial flow has been established through a network it is useful to know the extent to which the flow along any edge may be altered in either direction.

This is labelled as **excess capacity** and **back capacity** on each edge as shown. It may be possible to **augment** the flow through the network by making use of this flexibility along a path from the source vertex to the sink vertex. Such a path is called a **flow-augmenting path**.

A **saturated** edge is one where the flow is equal to the capacity.

An algorithm for finding the maximum flow through a network by augmentation is:

Step 1 Find an initial flow by inspection.

Step 2 Label the excess capacity and back capacity for each edge.

Step 3 Search for a flow-augmenting path. If one can be found then increase the flow along the path by the maximum amount that remains feasible.

Step 4 Repeat steps 2 and 3 until no flow-augmenting path may be found.

Multiple sources and sinks

A network may have several sources S_1, S_2, ... and several sinks T_1, T_2, ... but the methods described in this section can still be applied through the use of a **supersource** and a **supersink**. A supersource S is an extra node, added to the network, that acts as a source for each of the original sources. Similarly, a supersink T is a node, added to the network, that acts as a sink for each of the original sinks.

Example

The network below shows two sources S_1 and S_2 and two sinks T_1 and T_2. Find the maximum flow through the network and state any flow-augmenting paths.

Explain why the flow is maximal.

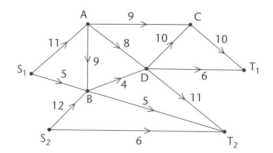

The first step is to add the supersource S and the supersink T to the network.

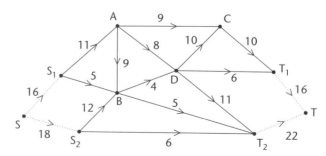

The usual convention is to use dotted lines for the edges linking S and T to the network.

The capacities on the edges SS_1 and SS_2 have been chosen to support the maximum possible flows from S_1 and S_2. Similarly, the capacities on the edges T_1T and T_2T have been chosen to support the maximum flows into T_1 and T_2. The diagram also shows a cut with capacity $11 + 4 + 5 + 6 = 26$.

A flow of 26 can be achieved as shown in the table.

Flow-augmenting path	Flow	Saturated edges
SS_1ACT_1T	9	S_1A, AC
SS_1ADT_1T	2	S_1A, DT_1
SS_2BDT_1T	4	BD, DT_1
SS_2BT_2T	5	BT_2
SS_2T_2T	6	S_2T_2
Total	26	

The flows on the flow-augmenting paths are shown circled on the network below. The flow of 26 is maximal since it equals the capacity of the cut (maximum flow minimum cut theorem).

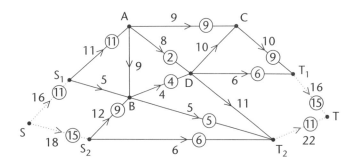

Note that each of the edges on the minimum cut is saturated. This will always be the case.

The final stage is to remove the supersource and supersink, along with the extra arcs, to show the maximum flow pattern for the original network.

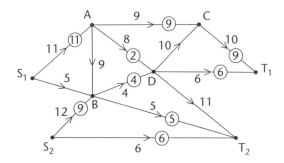

Progress check

1 Find the maximum flow through this network. Verify your answer using the maximum flow-minimum cut theorem.

2 Use a supersource and supersink to find the maximum flow through this network.
   Establish the flow through B in this case.

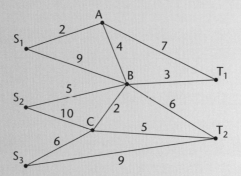

4.3 Critical path analysis

After studying this section you should be able to:

- construct an activity network from a given precedence table including the use of dummies where necessary
- use forward and backward scans to determine earliest and latest event times
- find the critical path in an activity network
- calculate the total float of an activity
- construct a Gantt chart for the purpose of scheduling
- draw a resource histogram and use resource levelling

The process of representing a complex project by a network and using it to identify the most efficient way to manage its completion is called **critical path analysis**.

Activity networks

A complex project may be divided into a number of smaller parts called **activities**. The completion of one or more activities is called an **event**.

Activities often rely on the completion of others before they can be started.

The relationship between these activities can be represented in a **precedence table**, sometimes called a **dependency table**.

In the precedence table shown on the right, the figures in brackets represent the **duration** of each activity, i.e. the time required, in hours, for its completion.

A precedence table can be used to produce an **activity network**. In the network, activities are represented by arcs and events are represented by vertices.

Activity	Depends on
A(3)	–
B(5)	–
C(2)	A
D(3)	A
E(3)	B, D
F(5)	C, E
G(1)	C
H(2)	F, G

The vertices are numbered from 0 at the **start vertex** and finishing at the **terminal vertex**.

The direction of the arrows shows the order in which the activities must be completed.

There must only be *one* activity between each pair of events in the network. The notation (i, j) is used to represent the activity between events i and j.

A **dummy activity** is one that has zero duration. A dummy is needed in this network to show that G depends on C whereas F depends on C *and* E.

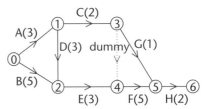

A dummy is shown with a dotted line. Its direction is important in defining dependency. In this case it shows that F depends on C not that G depends on E.

Earliest event times

The **earliest event time** for vertex i is denoted by e_i and represents the earliest time of arrival at event i with all dependent activities completed. These times are calculated using a **forward scan** from the start vertex to the terminal vertex.

Latest event times

The **latest event time** for vertex i is denoted by l_i and represents the latest time that event i may be left without extending the time for the project. These times are calculated using a **backward scan** from the terminal vertex back to the start vertex. The **critical path** is the longest path through the network. The activities on this path are the **critical activities**. If any critical activity is delayed then this will increase the time needed to complete the project. The events on the critical path are the **critical events** and for each of these $e_i = l_i$.

It is useful to add the information about earliest and latest times to the network. The critical path is then easily identified.

Notice that B(5) lies between two critical events but it is not a critical activity.

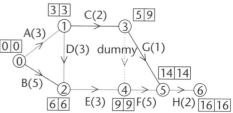

> The total float of any critical activity is always zero.

The **total float of an activity** is the maximum time that the activity may be delayed without affecting the length of the critical path. It is given by:

latest finish time – earliest start time – duration of the activity.

Scheduling

The process of allocating activities to workers for completion, within all of the constraints of the project, is known as **scheduling**.

The information regarding earliest and latest times for each activity is crucial when constructing a schedule. This information may be presented as a table or as a chart.

> Typically, the purpose of scheduling is to determine the number of workers needed to complete the project in a given time, or to determine the minimum time required for a given number of workers to complete the project.

Activity	Duration	Start Earliest	Start Latest	Finish Earliest	Finish Latest	Float
A(0, 1)	3	0	0	3	3	0
B(0, 2)	5	0	1	5	6	1
C(1, 3)	2	3	7	5	9	4
D(1, 2)	3	3	3	6	6	0
E(2, 4)	3	6	6	9	9	0
F(4, 5)	5	9	9	14	14	0
G(3, 5)	1	5	13	6	14	8
H(5, 6)	2	14	14	16	16	0

The critical activities are shown along one line.

> The diagram shown here is known as a cascade chart or Gantt chart.

The diagram illustrates the degree of flexibility in starting activities B, C and G. Remember that G cannot be started until C has been completed.

Resource histograms and resource levelling

A **resource histogram** shows the demand on resources over the time taken to complete a project.

The example below shows how the number of workers needed varies from day to day over a period of 14 days.

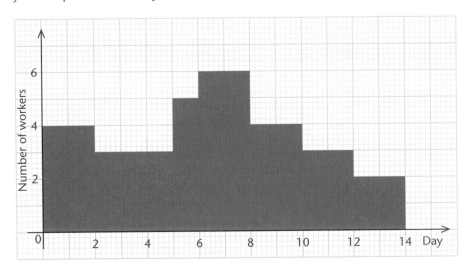

The maximum number of workers needed is 6 for days 7 and 8.

The diagram below shows the same resource histogram but includes details of the duration for each activity and the number of workers. Activity F, for example, lasts for 3 days and requires 2 workers. The critical activities are A, C, D, E and G. These are put in first and the rest of the diagram is built in layers on top.

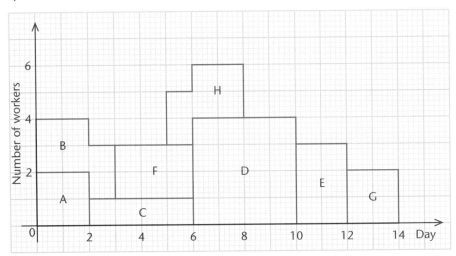

Once the resource histogram has been drawn, you may be asked to use **resource levelling** to work out the effect on the schedule of having a reduction in resources. For example, if the maximum number of workers available at any time is 5 then we can see from the diagram that this presents a problem for the completion of activity H.

Resource levelling involves selecting one or more activities to be delayed by an amount of time that allows the demand on resources to be brought down to the required level. The consequences of each delay must then be taken into account to determine the amount of extra time required.

In this example, delaying the start of activity D by 2 days allows activity H to be completed. Activities E and G must then also be delayed by 2 days and so the whole project requires 2 more days for completion.

Progress check

1 Determine the critical activities and the length of the critical path for this network.

2 Here is the activity network used in the example on page 118.

The number of workers required for each activity is shown in the table.

A	B	C	D	E	F	G	H
2	3	4	1	2	3	2	2

Assume that all workers are equally skilled and that once an activity has been started it must be completed without a break. Assume also that each activity starts at its earliest time.

(a) Draw the resource histogram for the project assuming that the number of workers available is unlimited.

(b) Given that the maximum number of workers available at any time is 5, use resource levelling to find the extra time needed to complete the project.

1 A, D, G and J, 21

2 (a)

(b) Delay the start of C by 2 days.
Delay the start of G by 2 days.
Delay the start of E, F and H by 1 day.
The extra time needed is 1 day.

4.4 Matching and allocation

After studying this section you should be able to:

- *use bipartite graphs to model matchings*
- *understand the conditions for matchings to be maximal or complete*
- *apply the maximum matching algorithm*

LEARNING SUMMARY

Matchings and graphs

A **bipartite graph** is a graph in which the vertices are divided into two sets such that no pair of vertices in the same set is connected by an edge.

In this case, the two sets are {A, B, C} and {p, q, r, s}.

Some vertices in a bipartite graph may not be connected to another vertex.

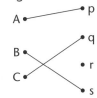

A **matching** between two sets may be represented by a bipartite graph in which there is at most one edge connecting a pair of vertices.

A **maximal matching** is a matching which has the maximum number of edges. This occurs when every vertex in one of the sets is connected to a vertex in the other set. The bipartite graph shown above represents a maximal matching.

A **complete matching** is a matching in which every vertex is connected to another vertex. This can only occur when the two sets contain the same number of vertices.

The matching improvement algorithm

Figure 1 is a bipartite graph showing the possible connections between two sets. It does not represent a matching because some vertices have more than one connection.
Figure 2 is a bipartite graph representing an initial matching.
An initial matching may be improved by increasing the number of connections. This is the purpose of the **matching improvement algorithm**.

Figure 1   Figure 2

> In an alternating path, the edges alternate between those from the bipartite graph that are not in the initial matching and those that are.

Step 1 In the initial matching, start from a vertex that is not connected and look for an **alternating path** to a vertex in the other set that is not connected.

Alternating path   Maximal matching

Step 2 Each edge on the alternating path, not included in the initial matching, is now included and each edge originally included is removed.

Step 3 Repeat steps 1 and 2 using the latest matching in place of the initial matching until no further alternating paths can be found.

Allocation problems

In an allocation (or assignment) problem, the object is to set up a matching that will optimise a particular objective, such as minimising a cost or maximising a profit. If we regard one set of objects as agents and the other set as tasks, then the underlying assumptions are that:

- Any agent may be assigned to any task (unlike the matching problems).
- Each agent is assigned at most one task and each task has at most one agent assigned to it.

When the number of agents is equal to the number of tasks, the problem is said to be **balanced**. If the number of agents exceeds the number of tasks, or vice versa, then the problem is **unbalanced**. The technique for solving an unbalanced problem involves introducing a **dummy** agent or task which then allows the solution to proceed as for the balanced case. In the solution of an unbalanced problem, a task assigned to a dummy agent isn't actually completed and an agent assigned to a dummy task doesn't actually have a task to carry out.

The opportunity cost matrix

The table shows the time in hours for three workers A, B and C to complete three separate tasks.

		Task		
		1	2	3
Worker	A	9	11	7
	B	8	10	9
	C	7	9	8

The **cost matrix** shows only the figures in hours. In this case, the 'cost' is measured in time.

9	11	7
8	10	9
7	9	8

The **opportunity cost matrix** is found from the cost matrix by:

1. Subtracting the smallest value in each row from every other value in that row.
2. Subtracting the smallest value in each column from every other value in that column.

9	11	7
8	10	9
7	9	8

$\xrightarrow{1}$

2	4	0
0	2	1
0	2	1

$\downarrow 2$

2	2	0
0	0	1
0	0	1

The Hungarian algorithm

The **Hungarian algorithm** is used to find the allocation that optimises the given objective.

Step 1 Find the opportunity cost matrix.

Step 2 Test for optimality. If true then make the allocation and stop.

Step 3 Revise the opportunity cost matrix and repeat step 2.

The method used to test for optimality is to find the minimum number of horizontal and/or vertical lines needed to cover all of the zeros in the opportunity cost matrix. If this number = n where the matrix has n rows and n columns, then the matrix will give an optimal allocation.

In the example, the number of lines needed = 3.
The matrix has 3 rows and columns so it will give an optimal allocation.

The optimal allocation is now found by selecting positions of zeros in the opportunity cost matrix such that one zero is selected in each row and in each column.

One possibility is shown here along with the corresponding figures in the original cost matrix.

2	2	**0**
0	**0**	1
0	0	1

9	11	**7**
8	**10**	9
7	9	8

The allocation is A→3, B→2 and C→1 giving a total cost, in hours, of 7 + 10 + 7 = 24.

In this case, there is an alternative solution.

2	2	**0**
0	0	1
0	**0**	1

9	11	**7**
8	10	9
7	**9**	8

The allocation is A→3, B→1 and C→2 giving a total cost, in hours, of 7 + 8 + 9 = 24.

Revising the opportunity cost matrix

When the opportunity cost matrix does not give an optimal solution, a revised opportunity cost matrix is found as follows:

Step 1 Find the smallest number not covered by a line. Subtract this number from every number not covered by a line.

Step 2 Add this number to every number lying at the intersection of two lines.

For example, this opportunity cost matrix has 3 rows and columns, but its zeros can be covered with just 2 lines.

0	3	1
0	2	5
0	0	0

Applying the steps given above produces this revised opportunity cost matrix.

0	2	**0**
0	1	4
1	**0**	0

Maximising allocation problems

If the objective involves maximising a quantity, such as a profit, then the first step is to subtract every value in the original matrix from a fixed value. This fixed value may be taken to be the largest value in the matrix or any value larger than this. The Hungarian algorithm is now applied in the usual way to this transformed matrix.

Progress check

1 Starting with an initial matching in which 1 is connected to R and 2 is connected to P, use the improvement algorithm to establish a complete matching for this bipartite graph.

2 Given the cost matrix:

	1	2	3
A	3	3	4
B	8	5	9
C	2	1	4

Use the Hungarian algorithm to find the minimum total cost.

2 $4 + 5 + 2 = 11$
1 $1 - P, 2 - S, 3 - R, 4 - Q$

4.5 Dynamic programming

After studying this section you should be able to:

- *understand the concept of dynamic programming*
- *set up a dynamic programming tabulation and use it to solve a problem*

LEARNING SUMMARY

Dynamic programming

Dynamic programming is one of the techniques used to solve problems where some quantity is to be optimised. It involves identifying **stages** and all possible **states** and **actions** within a process. The stages are counted from the end of the process to the beginning.

Working backwards through the stages, decisions are made at each stage that optimise the quantity. Importantly, the calculations carried out must follow from the corresponding optimal value found at the previous stage. In effect, this is applying the **principle of optimality** which states that **any part of an optimal path is itself optimal**.

In the examination, information about the process and the quantity to be optimised may be given in the form of a network or a table.

Example

The network below shows the possible actions and corresponding anticipated profits for a company over the next five months. The figures represent the profit in units of ten thousand pounds, with a negative value indicating a loss due to investment.

(a) Use dynamic programming to maximise the expected profit over the five months.
(b) Give the sequence of vertices in order to achieve the maximum profit.

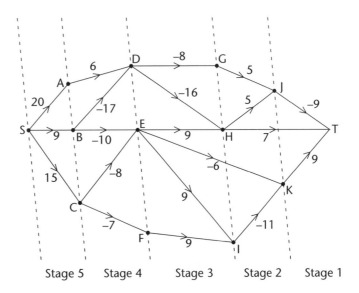

Again, in the examination, an insert is usually provided showing a table to be completed. The table on the next page is typical of the ones used and the coloured entries are typical of the information that would be given on the insert.

(a) An asterisk in the right-hand column is used to indicate the optimal value from the given state.

Stage	State	Action	Value	
1	J	JT	–9	*
	K	KT	9	*
2	G	GJ	5 + –9 = –4	*
	H	HJ	5 + –9 = –4	
		HT	7	*
	I	IK	–11 + 9 = –2	*
3	D	DG	–8 + –4 = –12	
		DH	–16 + 7 = –9	*
	E	EH	9 + 7 = 16	*
		EK	–6 + 9 = 3	
		EI	9 + –2 = 7	
	F	FI	9 + –2 = 7	*
4	A	AD	6 + –9 = –3	*
	B	BD	–17 + –9 = –26	
		BE	–10 + 16 = 6	*
	C	CE	–8 + 16 = 8	*
		CF	–7 + 7 = 0	
5	S	SA	20 + –3 = 17	
		SB	9 + 6 = 15	
		SC	15 + 8 = 23	*

JT and KT are the only possible paths from J and K, so they are automatically optimal.

The value for GJ is calculated as 5 (shown on GJ) plus –9 (the optimal value from J). This is the only value from G so it is automatically optimal.

There are two possible moves from H. The optimal value is 7.

There are three possible moves from E. The value for EH, for example, is 9 (shown on EH) plus 7 (the optimum value for H). This gives 16, which turns out to be the optimal value from E.

F is the last vertex to consider from Stage 3.

The three vertices to consider in Stage 4 are A, B and C.

The highest value found at the final stage is the overall maximum value. In this case this is 23 which represents 23 × £10 000.

The maximum expected profit over the next five months is £230 000.

(b) The sequence of vertices that maximises the profit is SCEHT.

Different types of problem

The dynamic programming technique may be used to solve various types of problem. The way in which the table is used remains the same, but the way in which the optimum value is selected may be different. To find the minimum distance through a network, for example, select the minimum value for each state rather than the maximum as in the example above.

In particular, two types that you need to be aware of are the **maximin** and **minimax**.

The optimal values for the maximin are found by first selecting the minimum value for each action and then choosing the maximum of these to represent the state.

To illustrate how the maximin works, here is part of a network and the corresponding part of a table.

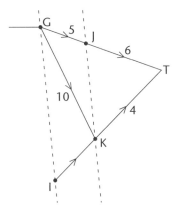

JT and KT are the only possible paths from J and K, so they are automatically optimal.

The value for GJ is the minimum of 5 (shown on the arc GJ) and 6 (the optimum from J at the previous stage).

The value for GK is the minimum of 10 (shown on the arc GJ) and 4 (the optimum from K at the previous stage).

Stage	State	Action	Value	
1	J	JT	6	*
	K	KT	4	*
2	G	GJ	Min(5, 6) = 5	*
		GK	Min(10, 4) = 4	

There are two values for the state G and the maximum of these (5) is shown with an asterisk as this is the optimum for G.

In a similar way, the optimal values for the minimax are found by first selecting the maximum value for each action and then choosing the minimum of these to represent the state.

The table shows the minimax values for the part of the network given above.

Stage	State	Action	Value	
1	J	JT	6	*
	K	KT	4	*
2	G	GJ	Max(5, 6) = 6	*
		GK	Max(10, 4) = 10	

Progress check

1 The diagram shows part of a network. Find the maximin value for the initial state A.

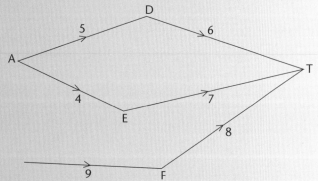

2 Find the minimax value for the initial state A in the diagram in question 1.

2 6

1 5

Sample questions and model answers

1

Ashley and Emma play a two-person zero-sum game in which they each consider three possible strategies. The table shows the pay-off matrix for Ashley.

		Emma		
		X	Y	Z
	A	5	−3	3
Ashley	B	4	1	−2
	C	−3	6	−4

(a) Find the play-safe strategy for each player and show that the game does not have a stable solution.

(b) Set up a linear programming problem to find Ashley's optimal mixed strategy. You are not required to solve it.

(a) The minimum row values are −3, −2 and −4 in Ashley's pay-off matrix. The maximum of these is −2 and so the play-safe strategy for Ashley is strategy B.

The minimum column values in Emma's pay-off matrix are −5, −6 and −3. The maximum of these is −3 and so the play-safe strategy for Emma is strategy Z.

		Emma		
		X	Y	Z
	A	5	−3	3
Ashley	B	4	1	−2
	C	−3	6	−4

		Emma		
		X	Y	Z
	A	−5	3	−3
Ashley	B	−4	−1	2
	C	3	−6	4

The values used to determine the play-safe strategies do not have the corresponding positions in the pay-off matrices so there is no stable solution. Alternatively, −3 + −2 ≠ 0 and so, again, there is no stable solution.

Sample questions and model answers (continued)

(b) Adding 5 to each of the figures in Ashley's pay-off matrix gives:

> The values need to be made positive by adding a fixed value to each one.

10	2	8
9	6	3
2	11	1

Using a, b and c to represent the probability that Ashley uses strategy A, B or C respectively, the linear programming problem is formulated as:

Maximise P subject to $a + b + c \leqslant 1$ where $a \geqslant 0$, $b \geqslant 0$, $c \geqslant 0$ and

$$P - 10a - 9b - 2c \leqslant 0$$
$$P - 2a - 6b - 11c \leqslant 0$$
$$P - 8a - 3b - c \leqslant 0$$

and P represents the value of the game.

2

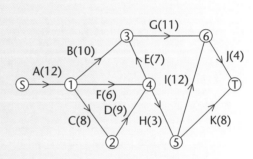

The diagram is an activity network for a project.

The figures in brackets represent the time in days to complete each activity.

The start vertex is S and the terminal vertex is T.

(a) Label each vertex with the earliest and latest time for the event. Give the length of the critical path.

(b) State which events are critical and give the critical path.

> Use a forward scan to find the earliest event times.

> Use a backward scan to find the latest event times.

(a) The length of the critical path is 51 days.

(b) The critical events are S, 1, 2, 3, 4, 6 and T.

The critical path is A, C, D, E, G and J.

> The length of the critical path is given by the earliest (and latest) event time at the terminal vertex.

Practice examination questions

1 A and B are the players in a two-person zero-sum game. Each may use one of two strategies when a game is played. The pay-off matrix for A is shown below.

Player A		Player B	
		Y	Z
	W	4	8
	X	6	3

(a) Find the best mixed strategy for each player.

(b) Find the value of the game.

2 Richard and Judy play a zero-sum game. The pay-off matrix for Richard is shown in the table.

Richard		Judy		
		X	Y	Z
	A	4	3	−4
	B	−3	6	8
	C	2	1	−6

(a) Find the play-safe strategy for each player.

(b) Show that there is no saddle point for the game.

(c) Explain why strategy C will not be part of Richard's mixed strategy.

(d) Find Richard's optimal mixed strategy.

Practice examination questions *(continued)*

3 The table shows the names of five employees of a company and the days when they are each available to work a late shift. The days that have been highlighted represent a first attempt at matching the employees to the available days to make a roster.

Name	Days available
Abbie	**Mon**, Tue, Fri
Ben	Mon, **Wed**, Thu
Colin	Mon, **Fri**
Dave	Fri, Wed
Emma	**Thu**, Fri

(a) Use a bipartite graph to represent the relationship between employees and days available.

(b) Use a second bipartite graph to represent an initial matching based on the days highlighted.

(c) Implement a matching improvement algorithm to obtain a complete matching. Indicate the alternating path used.

4 An expedition is planned across a desert from S to T. The network shows the possible routes and distances in km between places where water and supplies are available. The optimal route is the minimax route that minimises the maximum distance between supply points.

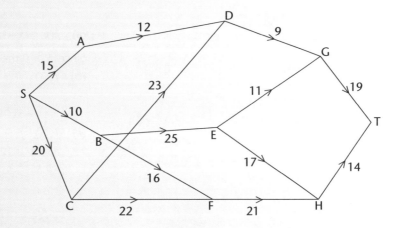

By completing the following table, or otherwise, use dynamic programming, working backwards from T, to find the minimax route and give the maximum distance between supply points on the route.

Stage	State	Action	Value	
1	G	GT	19	*
	H	HT	14	*
2	D	DG	Max(9, 19) = 19	
	E	EH		
	F	FH		
3	A	AD		
	B	BE		
		BF		
4				

Practice examination answers

1

$$\frac{1}{(1+x)^2} = (1+x)^{-2}$$

$$= 1 + (-2)x + \frac{(-2)(-3)}{2!}x^2 + \ldots$$

$$= 1 - 2x + 3x^2 + \ldots \quad |x| < 1$$

$$\sqrt{4+x} = \sqrt{4\left(1 + \frac{x}{4}\right)}$$

$$= 4^{\frac{1}{2}}\left(1 + \frac{x}{4}\right)^{\frac{1}{2}}$$

$$= 2\left(1 + \frac{x}{4}\right)^{\frac{1}{2}}$$

$$= 2\left(1 + \left(\tfrac{1}{2}\right)\left(\frac{x}{4}\right) + \frac{\left(\tfrac{1}{2}\right)\left(-\tfrac{1}{2}\right)}{2!}\left(\frac{x}{4}\right)^2 + \ldots\right)$$

$$= 2 + \frac{x}{4} - \frac{x^2}{64} + \ldots \quad |x| < 4$$

$$\therefore \quad f(x) = 1 - 2x + 3x^2 + 2 + \frac{x}{4} - \frac{x^2}{64} + \ldots$$

$$= 3 - 1\tfrac{3}{4}x + 2\tfrac{63}{64}x^2 + \ldots \quad |x| < 1$$

Ignoring terms in x^3 and higher, $f(x) \approx a + bx + cx^2$, with $a = 3$, $b = -1\tfrac{3}{4}$, $c = 2\tfrac{63}{64}$.

The restriction is $|x| < 1$.

2 (a)

$$y = \cos^2 x \sin x$$

$$\frac{dy}{dx} = \cos^2 x(\cos x) + \sin x(-2\sin x \cos x)$$

$$= \cos^3 x - 2\cos x \sin^2 x$$

$$= \cos^3 x - 2\cos x(1 - \cos^2 x)$$

$$= \cos^3 x - 2\cos x + 2\cos^3 x$$

$$= 3\cos^3 x - 2\cos x$$

(b)

$$u = \cos x \Rightarrow \frac{du}{dx} = -\sin x$$

$$\frac{dx}{du} = -\frac{1}{\sin x} \Rightarrow \sin x \frac{dx}{du} = -1$$

$$\int \cos^2 x \sin x \, dx$$

$$= \int \cos^2 x \sin x \frac{dx}{du} du$$

$$= \int (-u^2) du$$

$$= -\tfrac{1}{3}u^3 + c$$

$$= -\tfrac{1}{3}\cos^3 x + c$$

(This can be also be done by direct recognition, see page 43.)

3

$$\frac{dy}{dx} = \frac{2x(x+1) - 1(x^2 - 4)}{(x+1)^2}$$

$$= \frac{x^2 + 2x + 4}{(x+1)^2}$$

When $x = 2$, $\dfrac{dy}{dx} = \dfrac{4}{3}$

$\therefore$ gradient of normal $= -\tfrac{3}{4}$

Equation of normal at $(2, 0)$:

$$y - 0 = -\tfrac{3}{4}(x - 2)$$

$$4y = -3x + 6$$

$$3x + 4y - 6 = 0$$

$\therefore \quad a = 3$, $b = 4$ and $c = -6$.

4 (a) $g(x) = (x+1)(x-1) = x^2 - 1$.

Since $x^2 \geqslant 0$ the range of g is $x \geqslant -1$.

(b) $fg(x) = f(x^2 - 1)$

$$= 5(x^2 - 1) - 7$$

$$= 5x^2 - 12.$$

(c) $fg(x) = f(x)$

$$\Rightarrow 5x^2 - 12 = 5x - 7$$

$$\Rightarrow 5x^2 - 5x - 5 = 0$$

$$\Rightarrow x^2 - x - 1 = 0$$

$$\Rightarrow x = \frac{1 \pm \sqrt{5}}{2}.$$

5 (a) $R\sin(x + a) \equiv 15\sin x + 8\cos x$

$\Rightarrow R\cos a = 15$ and $R\sin a = 8$ (a is acute)

$\Rightarrow R = \sqrt{15^2 + 8^2} = 17$

$\tan a = \tfrac{8}{15} \Rightarrow a = 28.07°$

So $15\sin x + 8\cos x \equiv 17\sin(x + 28.07°)$.

(b) Max value is 17.

(c) $15\sin x + 8\cos x = 12$

$\Rightarrow 17\sin(x + 28.07°) = 12.$

$\sin^{-1}(\tfrac{12}{17}) = 44.90°$, $180° - 44.90° = 135.10°$

$x + 28.07 = 44.90° \Rightarrow x = 16.8°$

$x + 28.07 = 135.10° \Rightarrow x = 107.0°$

Core 3 and Core 4 (Pure Mathematics) *(continued)*

6 $x\dfrac{dy}{dx} = y + yx \Rightarrow x\dfrac{dy}{dx} = y(1+x)$

Separating the variables:

$\dfrac{1}{y}\dfrac{dy}{dx} = \dfrac{1+x}{x}$

$\dfrac{1}{y}\dfrac{dy}{dx} = \dfrac{1}{x} + 1$

$\displaystyle\int \dfrac{1}{y}\,dy = \int \left(\dfrac{1}{x}+1\right)dx$

$\therefore$ $\ln y = \ln x + x + c$

When $x = 2$, $y = 4$,

$\Rightarrow$ $\ln 4 = \ln 2 + 2 + c$

$\Rightarrow$ $c = \ln 2 - 2$ $(\ln 4 = 2\ln 2)$

So $\ln y = \ln x + x + \ln 2 - 2$

$\ln\left(\dfrac{y}{2x}\right) = x - 2$

$\dfrac{y}{2x} = e^{x-2}$

$y = 2xe^{x-2}$

7 (a) (i) $y = (4x-3)^9$

$\Rightarrow \dfrac{dy}{dx} = 9(4x-3)^8 \times 4$

$\Rightarrow \dfrac{dy}{dx} = 36(4x-3)^8.$

(ii) $y = \ln(5-2x)$

$\dfrac{dy}{dx} = \dfrac{1}{5-2x} \times (-2)$

$= \dfrac{-2}{5-2x}$ $\left(\text{you can write this as } \dfrac{2}{2x-5}\right).$

(b) (i) $\displaystyle\int (4x-3)\,dx = \dfrac{1}{36}(4x-3)^9 + C.$

(ii) $\displaystyle\int_3^4 \dfrac{1}{5-2x}\,dx = -\dfrac{1}{2}\int_3^4 \dfrac{-2}{5-2x}\,dx$

$= -\dfrac{1}{2}\,[\ln|5-2x|]_3^4$

$= -\dfrac{1}{2}\,(\ln|-3| - \ln|-1|)$

$= -\dfrac{1}{2}\ln 3.$

8 (a) $\displaystyle\int_0^1 xe^{-2x}\,dx = \left[x\left(-\tfrac{1}{2}e^{-2x}\right)\right]_0^1 - \int_0^1 \left(-\tfrac{1}{2}e^{-2x}\right)dx$

$= -\tfrac{1}{2}e^{-2} + \tfrac{1}{2}\displaystyle\int_0^1 e^{-2x}\,dx$

$= -\tfrac{1}{2}e^{-2} + \tfrac{1}{2}\left[\left(-\tfrac{1}{2}e^{-2x}\right)\right]_0^1$

$= -\tfrac{1}{2}e^{-2} + \tfrac{1}{2}\left(-\tfrac{1}{2}e^{-2} - \left(-\tfrac{1}{2}\right)\right)$

$= -\tfrac{3}{4}e^{-2} + \tfrac{1}{4}$

(b) (i) $y = xe^{-x}$

$\dfrac{dy}{dx} = x(-e^{-x}) + e^{-x} = e^{-x}(1-x)$

$\dfrac{dy}{dx} = 0$ when $x = 1$

When $x = 1$, $y = e^{-1} \Rightarrow A$ is point $(1, e^{-1})$.

(ii) $V = \pi\displaystyle\int_0^1 y^2\,dx = \pi\int_0^1 x^2 e^{-2x}\,dx$

$= \pi\left(\left[x^2\left(-\tfrac{1}{2}e^{-2x}\right)\right]_0^1 - \int_0^1 -\tfrac{1}{2}e^{-2x}(2x)\,dx\right)$

$= \pi\left(-\tfrac{1}{2}e^{-2} + \int_0^1 xe^{-2x}\,dx\right)$

$= \pi\left(-\tfrac{1}{2}e^{-2} - \tfrac{3}{4}e^{-2} + \tfrac{1}{4}\right)$ (using part (a))

$= \tfrac{\pi}{4}\left(1 - 5e^{-2}\right)$

9 $\dfrac{dP}{dt} \propto P \Rightarrow \dfrac{dP}{dt} = kP$, where $k > 0$.

Separating the variables

$\displaystyle\int \dfrac{1}{P}\,dP = \int k\,dt$

$\Rightarrow$ $\ln P = kt + c$

$\Rightarrow$ $P = e^{kt+c}$

$\Rightarrow$ $P = Ae^{kt}$ (where $A = e^c$)

$t = 0$, $P = 500 \Rightarrow 500 = A$

$t = 10$, $P = 1000 \Rightarrow 1000 = 500e^{10k}$

$\therefore$ $2 = e^{10k}$

$\Rightarrow$ $10k = \ln 2$

$\Rightarrow$ $k = 0.1\ln 2$

$\therefore$ $P = 500e^{(0.1\ln 2)t}$

When $t = 20$,

$P = 500e^{0.1\ln 2 \times 20} = 500 \times 4 = 2000$

So the population is 2000 when $t = 20$.

Core 3 and Core 4 (Pure Mathematics) (continued)

10 (a)
$$x^3 + xy + y^2 = 7$$
$$3x^2 + x\frac{dy}{dx} + y + 2y\frac{dy}{dx} = 0$$

At (1, 2)
$$3 + \frac{dy}{dx} + 2 + 4\frac{dy}{dx} = 0$$
$$\Rightarrow 5\frac{dy}{dx} + 5 = 0 \text{ i.e. } \frac{dy}{dx} = -1$$

At (1, 2) gradient of normal = 1

Equation of normal is
$$y - 2 = 1(x - 1)$$
$$\text{i.e. } y = x + 1$$

(b) $x = t^2, \ y = t + \dfrac{1}{t}$.

$$\frac{dx}{dt} = 2t$$
$$\frac{dy}{dt} = 1 - \frac{1}{t^2} = \frac{t^2 - 1}{t^2}$$
$$\frac{dy}{dx} = \frac{dy}{dt} \times \frac{dt}{dx} = \frac{t^2 - 1}{2t^3}$$
$$\frac{dy}{dx} = 0 \text{ when } t^2 - 1 = 0 \Rightarrow t = \pm 1$$

When $t = 1$, coordinates are (1, 2).
When $t = -1$, coordinates are (1, −2).
The stationary points are at (1, 2) and (1, −2).

11 (a) $\sin 2x = 0$ when $2x = 0, \pi, 2\pi, \ldots$
i.e. when $x = 0, \frac{1}{2}\pi, \pi, \ldots$
$$\therefore a = \tfrac{1}{2}\pi$$

(b) $A = \displaystyle\int_0^{\frac{1}{2}\pi} y\,dx = \int_0^{\frac{1}{2}\pi} \sin 2x\,dx$

$$= -\tfrac{1}{2}\Big[\cos 2x\Big]_0^{\frac{1}{2}\pi}$$
$$= -\tfrac{1}{2}(-1 - 1)$$
$$= 1$$

Area of R is 1 square unit.

(c) $V = \pi \displaystyle\int_0^{\frac{1}{2}\pi} y^2\,dx$

$$= \pi \int_0^{\frac{1}{2}\pi} \sin^2 2x\,dx$$
$$= \tfrac{1}{2}\pi \int_0^{\frac{1}{2}\pi} (1 - \cos 4x)\,dx$$
$$= \tfrac{1}{2}\pi\Big[x - \tfrac{1}{4}\sin 4x\Big]_0^{\frac{1}{2}\pi}$$
$$= \tfrac{1}{2}\pi(\tfrac{1}{2}\pi - 0 - 0)$$
$$= \tfrac{1}{4}\pi^2$$

The volume generated is $\tfrac{1}{4}\pi^2$ cubic units.

12 (a) Referring to origin (0, 0)
$$\mathbf{a} = 2\mathbf{i} + 2\mathbf{j} \text{ and } \mathbf{b} = 2\mathbf{i} + 3\mathbf{j} + \mathbf{k}$$

Also $\mathbf{b} - \mathbf{a} = \mathbf{j} + \mathbf{k}$

A vector equation for AB is

$\mathbf{r} = \mathbf{a} + \mu(\mathbf{b} - \mathbf{a})$ where μ is a scalar parameter

$$\therefore \quad \mathbf{r} = 2\mathbf{i} + 2\mathbf{j} + \mu(\mathbf{j} + \mathbf{k})$$

(b) If lines intersect, there are unique values for μ and λ for which

$$2\mathbf{i} + 2\mathbf{j} + \mu(\mathbf{j} + \mathbf{k}) = 2\mathbf{j} + \mathbf{k} + \lambda(\mathbf{i} + \mathbf{k})$$

Equating coefficients of $\mathbf{i}$, $\mathbf{j}$ and $\mathbf{k}$, in turn gives:

$$2 = \lambda$$
$$2 + \mu = 2$$
$$\mu = 1 + \lambda$$

There are no values of μ and λ that satisfy all three equations, so the lines have no common point.

(c) The angle between the lines is the angle between their direction vectors.

$$|\mathbf{j} + \mathbf{k}| = \sqrt{2} \text{ and } |\mathbf{i} + \mathbf{k}| = \sqrt{2}$$
$$(\mathbf{j} + \mathbf{k}).(\mathbf{i} + \mathbf{k}) = 1$$

Let the angle be θ, then

$$\cos\theta = \frac{1}{\sqrt{2}\sqrt{2}} = 0.5$$
$$\Rightarrow \qquad \theta = 60°$$

Probability and Statistics 2

1 If $\mu = 20$, $\bar{X} \sim N(20, \frac{16}{10})$,

i.e. $\bar{X} \sim N(20, 1.6)$.

Carrying out a two-tailed test at 5% level, reject H_0 if $z < -1.96$ where

$$z = \frac{17.2 - 20}{\sqrt{1.6}} = -2.213\ldots$$

Since $z < -1.96$, reject H_0.

There is evidence that μ is not 20.

2 (a) If X is number of people with the health problem, then $X \sim B(80, 0.01)$. n is large and $np < 5$, so $X \sim Po(np)$, i.e. $X \sim Po(0.8)$ approximately.

$$P(X < 3) = e^{-0.8} + 0.8\, e^{-0.8} + \frac{0.8^2}{2!}\, e^{-0.8}$$

$$= 0.953 \text{ (3 s.f.)}$$

(b) If X is number of people with the health problem in a sample of size n, $X \sim B(n, 0.01)$. If $n > 50$, $X \sim P(0.01n)$ approximately.

$$P(X \geq 1) = 1 - P(X = 0) = 1 - e^{-0.01n}$$

So $P(X \geq 1) > 0.95 \Rightarrow 1 - e^{-0.01n} > 0.95$ i.e. $e^{-0.01n} < 0.05$

By trial and improvement, $e^{-2.99} = 0.0502 > 0.05$, $e^{-3} = 0.049 < 0.05$, so $0.01n = 3 \Rightarrow n = 300$. (Log theory could be used here.) The minimum number in the sample should be 300.

3 (a) $k = \dfrac{1}{10}$

(b) By symmetry, $E(X) = \frac{1}{2}(2 + 12) = 7$

$$E(X^2) = \frac{1}{10} \int_2^{12} x^2 \, dx = \frac{1}{10}\left[\frac{x^3}{3}\right]_2^{12} = 57\tfrac{1}{3}$$

$$Var(X) = 57\tfrac{1}{3} - 7^2 = 8\tfrac{1}{3}$$

Standard deviation $= \sqrt{8\tfrac{1}{3}} = 2.887$ (3 d.p.)

$P(7 - 2.887 < X < 7 + 2.887)$

$= P(4.113 < X < 9.887)$

$= 0.1 \times (9.887 - 4.113)$

$= 0.58$ (2 d.p.)

4 (a) $X \sim N(3.5, 0.8^2)$

$$P(X < 3.0) = P\!\left(Z < \frac{3.3 - 3.5}{0.8}\right)$$

$$= P(Z < -0.25)$$

$$= 1 - \Phi(0.25)$$

$$= 1 - 0.5987$$

$$= 0.4013$$

(b) $$P(X > 3.25) = 0.95$$

$$P\!\left(Z > \frac{3.25 - 3.5}{\sigma}\right) = 0.95$$

Now $P(Z > z) = 0.95 \Rightarrow z = -1.645$

so $$\frac{3.25 - 3.5}{\sigma} = -1.645$$

$$3.25 - 3.5 = -1.645\sigma$$

$$-0.25 = -1.645\sigma$$

$$\sigma = \frac{-0.25}{-1.645} = 0.151\ldots = 0.15 \text{ (2 d.p.)}$$

5 For the binomial distribution,

$E(X) = np = 10 \times 0.4 = 4$

$Var(X) = npq = 4 \times 0.6 = 2.4$

By the central limit theorem, since the sample size is large,

$$\bar{X} \sim N\!\left(4, \frac{2.4}{60}\right), \text{ i.e. } \bar{X} \sim N(4, 0.04)$$

$$P(\bar{X} < 3.5) = P\!\left(Z < \frac{3.5 - 4}{\sqrt{0.04}}\right)$$

$$= P(Z < -2.5)$$

$$= 0.0062$$

Probability and Statistics 2 *(continued)*

6 Let X be the number of trees with the infestation. Assuming that trees are infested independently, with probability p, $X \sim B(n, p)$

$H_0: p = 0.35$

$H_1: p > 0.35$

(a) If $p = 0.35$, $X \sim B(10, 0.35)$

At the 10% level,

H_0 is rejected if $P(X \geqslant 6) < 0.1$

Using cumulative binomial tables,

$P(X \geqslant 6) = 1 - P(X \leqslant 5)$

$\qquad = 1 - 0.9051 = 0.0949$

Since $P(X \geqslant 6) < 0.1$, H_0 is rejected.

There is evidence, at the 10% level, that the percentage of trees that are infested is greater than 35%, indicating that the trees should be felled.

(b) If $p = 0.35$, $X \sim B(30, 0.35)$

n is large such that $np = 10.5 > 5$ and

$nq = 19.5 > 5$ so use normal approximation.

$npq = 10.5 \times 0.65 = 6.825$

so $X \sim N(10.5, 6.825)$ approximately.

At the 10% level, H_0 is rejected if $z > 1.282$

Test $x = 14$.

Applying a continuity correction,

$z = \dfrac{13.5 - 10.5}{\sqrt{6.825}} = 1.148$

Since $z < 1.282$, H_0 is not rejected.

There is not enough evidence to say that the percentage of infected trees is greater than 35%. The trees should be treated with chemicals.

7 (a) $\displaystyle\int_{\text{all } x} f(x)\, dx = 1$

so $\quad 1 = k \displaystyle\int_0^1 (x - x^3)\, dx$

$\qquad = k \left[\dfrac{x^2}{2} - \dfrac{x^4}{4} \right]_0^1$

$\qquad = k \left(\dfrac{1}{2} - \dfrac{1}{4} \right)$

$\qquad = \dfrac{1}{4} k$

$\therefore \qquad k = 4$

(b) $\mu = E(X) = \displaystyle\int_0^1 x f(x)\, dx = 4 \int_0^1 x(x - x^3)\, dx$

$\qquad = 4 \displaystyle\int_0^1 (x^2 - x^4)\, dx$

$\qquad = 4 \left[\dfrac{x^3}{3} - \dfrac{x^5}{5} \right]_0^1$

$\qquad = 4 \left(\dfrac{1}{3} - \dfrac{1}{5} \right)$

$\qquad = \dfrac{8}{15}$

(c) $P(X < 0.5) = 4 \displaystyle\int_0^{0.5} (x - x^3)\, dx$

$\qquad = 4 \left[\dfrac{x^2}{2} - \dfrac{x^4}{4} \right]_0^{0.5}$

$\qquad = 0.4375$

(d) $E(X^2) = \displaystyle\int_0^1 x^2 f(x)\, dx = 4 \int_0^1 x^2(x - x^3)\, dx$

$\qquad = 4 \displaystyle\int_0^1 (x^3 - x^5)\, dx$

$\qquad = 4 \left[\dfrac{x^4}{4} - \dfrac{x^6}{6} \right]_0^1$

$\qquad = \dfrac{1}{3}$

$\text{Var}(X) = E(X^2) - \mu^2$

$\qquad = \dfrac{1}{3} - \left(\dfrac{8}{15} \right)^2$

$\qquad = 0.0489 \text{ (3 s.f.)}$

Probability and Statistics 2 *(continued)*

8 (a) $P(X < 16) = P(X \leqslant 15) = 0.566$

(b) Let the number in stock be k.
We require $P(X > k) < 0.05 \Rightarrow P(X \leqslant k) \geqslant 0.95$

From tables, $P(X \leqslant 22) = 0.9418 < 0.95$

$P(X \leqslant 23) = 0.9633 > 0.95$

So there should be 23 copies in stock to meet
the demand.

Mechanics 2

1 (a)   Using $y = x \tan\theta - \dfrac{gx^2}{2v^2}(1 + \tan^2\theta)$

gives $2.5 = 20 \tan\theta - \dfrac{9.8 \times 20^2}{2 \times 25^2}(1 + \tan^2\theta)$

$3.136 \tan^2\theta - 20 \tan\theta + 5.636 = 0$

$\tan\theta = 6.082\ldots$ or $\tan\theta = 0.2954\ldots$

$\tan^{-1}(0.2954\ldots) = 16.5°$

so a possible angle of projection is $16.5°$.

(b)   Vertically: using $s = ut + \frac{1}{2}at^2$ gives

$-0.5 = 25 \sin 16.5°\, t - 4.9t^2$

$4.9t^2 - 25 \sin 16.5°\, t - 0.5 = 0$

$t = 1.516\ldots$ or $t = -0.067\ldots$

since $t > 0$

$d = 25 \cos 16.5° \times 1.516 - 20$

The distance is 16.3 m to 3 s.f.

2 (a)   3 cm (by symmetry).

(b) $\bar{y} = \dfrac{24 \times 2 - 4 \times 3}{20} = 1.8$ cm

(c)

$\theta = \tan^{-1}\left(\frac{3}{1.8}\right) = 59.0°$

3 (a)   Area rectangle : area triangle $= 3 : 1$

so   mass of rectangle $= 0.6$ kg

mass of triangle $= 0.2$ kg

$\bar{x} = \dfrac{0.6 \times 7.5 + 0.2 \times 18\frac{1}{3}}{0.8} = 10.2$ cm

(b)

$15R_B = 10.2 \times 8 \Rightarrow R_B = 5.44$ N

$R_A = 8 - 5.44 = 2.56$ N

4 (a)

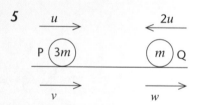

Conservation of momentum:

$0.9 \times 4 - 0.3 \times 2 = 0.9u + 0.3v$

so $3u + v = 10$ (1)

Newton's experimental law:

$v - u = \frac{1}{3} \times 6 = 2$ (2)

From (1) and (2) $4u = 8$

giving $u = 2$, $v = 4$.

The speeds of A and B after the collision are 2 ms^{-1} and 4 ms^{-1} respectively.

(b)   Impulse = change in momentum

$= 0.3 \times 4 - 0.3 \times (-2)$

$= 1.8$ Ns

(c)   Total KE before impact

$= \frac{1}{2} \times 0.9 \times 4^2 + \frac{1}{2} \times 0.3 \times 2^2 = 7.8$ J

Total KE after impact

$= \frac{1}{2} \times 0.9 \times 2^2 + \frac{1}{2} \times 0.3 \times 4^2 = 4.2$ J

Loss of mechanical energy $= 3.6$ J

5

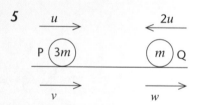

Conservation of momentum:

$3mu - 2mu = 3mv + mw$

giving: $3v + w = u$ (1)

Newton's experimental law:

$w - v = 3ue$ (2)

(1)–(2) gives $4v = u(1 - 3e)$

so $v = \dfrac{u}{4}(1 - 3e)$

$v < 0 \Rightarrow (1 - 3e) < 0$

$\Rightarrow e > \frac{1}{3}$

Mechanics 2 (continued)

6 (a) Before impact

After impact

Momentum before = momentum after

so $\quad 10mu = 2mv + 2mv$

$\Rightarrow 10u = 4v$

$\Rightarrow v = 2.5u$

(b) Impulse = change in momentum

$= 5mu - 0$

$= 5mu.$

7

vertically: $\quad R = mg$

horizontally: $\quad F = \dfrac{m \times 15^2}{40}$

$\mu = \dfrac{F}{R} = \dfrac{15^2}{40 \times 9.8} = 0.57$

8

(a) Magnitude of acceleration $= \omega^2 r$

$= 3^2 \times 0.3$

$= 2.7\ \text{ms}^{-2}$

(b) Using $F = ma$: $\quad T \cos\theta = 2 \times 2.7$

so: $\quad T \times \tfrac{3}{5} = 5.4$

giving: $\quad T = 9\ \text{N}$

(c)

Vertically: $\quad R + 9\sin\theta = 2 \times 9.8$

$R = 2 \times 9.8 - 9 \times 0.8 = 12.4$

$R = 12.4\ \text{N}$

9 (a) The tension is perpendicular to the direction of motion.

(b) The acceleration is not constant.

(c) Maximum speed occurs at lowest point.

Loss of PE = gain in KE

$mgl = \tfrac{1}{2}mv^2$

so, maximum speed is given by $v = \sqrt{2gl}$

10 (a)

Resolving parallel to the plane:

$F = 900 \times 9.8 \times \tfrac{1}{15} + 800$

$= 1388$

The driving force is 1388 N.

(b) $P = Fv = 1388 \times 20 = 27760$

Power $\qquad = 27.76\ \text{kW}$

(c) Using $F = ma$:

gives: $\quad 1388 - 800 = 900a \Rightarrow a = 0.65\ ms^{-2}.$

11 Momentum of ball before impact

$= 0.2 \times 20 = 4\ \text{Ns}$

Momentum of ball after impact

$= 0.2 \times 5 = 1\ \text{Ns}$

The magnitude of the impulse is given by:

$I^2 = 1^2 + 4^2 = 17 \Rightarrow I = \sqrt{17}$

$\tan\theta = \tfrac{1}{4} \Rightarrow \theta = 14.0°$

The impulse has magnitude $\sqrt{17}$ Ns at an angle of 14.0° above the horizontal.

Mechanics 2 (continued)

12 (a)

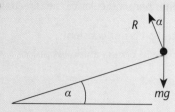

Vertically: $\quad\quad R \cos \alpha = mg$ $\quad\quad\quad$ (1)

Horizontally: $\quad\quad R \sin \alpha = \dfrac{m \times 20^2}{80}$ $\quad\quad\quad$ (2)

(2) ÷ (1) gives $\quad\quad \tan \alpha = \dfrac{20^2}{80 \times 9.8}$

$\quad\quad\quad\quad\quad\quad\quad\quad \Rightarrow \alpha = 27.0°$

(b)

vert: $R \cos 27° = mg + \mu R \sin 27°$ $\quad\quad\quad$ (3)

horiz: $R \sin 27° + \mu R \cos 27° = \dfrac{m \times 28^2}{80}$ $\quad\quad\quad$ (4)

(4) ÷ (3) gives:

$\dfrac{\sin 27° + \mu \cos 27°}{\cos 27° - \mu \sin 27°} = \dfrac{28^2}{80 \times 9.8}$

rearranging gives:

$\quad\quad \mu = \dfrac{28^2 \cos 27° - 80 \times 9.8 \sin 27°}{80 \times 9.8 \cos 27° + 28^2 \sin 27°}$

$\Rightarrow \mu = 0.325$ to 3 s.f.

13 (a) Horizontally:

$\quad T \sin 30° = \dfrac{0.2 \times 1.2^2}{0.4}$

$\quad\quad \Rightarrow T = 1.44$

The tension is 1.44 N.

Vertically:

$1.44 \cos 30° + R = 0.2 \times 9.8$

$\quad\quad\quad \Rightarrow R = 0.713$

The reaction is 0.713 N to 3 s.f.

(b) Vertically: the new tension is given by:

$T \cos 30° = 0.2 \times 9.8 \Rightarrow T = 2.263$

Horizontally: $\quad 2.263 \sin 30° = 0.2\omega^2 \times 0.4$

giving: $\quad\quad\quad\quad\quad\quad \omega = 3.76$

The minimum angular speed is 3.76 rad sec^{-1}.

Decision Mathematics 2

1 (a) Assume A uses Strategy W with probability p and strategy X with probability $(1 - p)$.

Expected gain for A is:

$4p + 6(1 - p)$ if B uses strategy Y.

$8p + 3(1 - p)$ if B uses strategy Z.

Optimal when

$4p + 6(1 - p) = 8p + 3(1 - p)$

$\Rightarrow -2p + 6 = 5p + 3$

$\Rightarrow p = \frac{3}{7}$

Assume B uses strategy Y with probability q and strategy Z with probability $(1 - q)$.

Expected loss for B is:

$4q + 8(1 - q)$ if A uses strategy W

$6q + 3(1 - q)$ if A uses strategy X.

Optimal when

$4q + 8(1 - q) = 6q + 3(1 - q)$

$\Rightarrow -4q + 8 = 3q + 3$

$\Rightarrow q = \frac{5}{7}$

The optimal strategy for A is to use strategy W with probability $\frac{3}{7}$ and strategy X with probability $\frac{4}{7}$.

The optimal strategy for B is to use strategy Y with probability $\frac{5}{7}$ and strategy Z with probability $\frac{2}{7}$.

(b) The value of the game is $5\frac{1}{7}$.

2 (a) The minimum row values are −4, −3 and −6. The maximum of these is −3 showing that the play-safe strategy for Richard is strategy B.

The minimum of the negatives of the column values are −4, −6 and −8. The maximum of these is −4 showing that the play-safe strategy for Judy is strategy X.

(b) $-3 + (-4) \neq 0$ so there is no saddle point.

(c) Comparing the row values shows that Richard's pay-offs are *always* less with strategy C than with strategy A. It follows that strategy C will not be a part of his mixed strategy.

(d) Assume Richard uses strategy A with probability p and strategy B with probability $(1 - p)$.

The expected gains for Richard based on the strategy used by Judy are:

X: $4p - 3(1 - p) = 7p - 3$

Y: $3p + 6(1 - p) = -3p + 6$

Z: $-4p + 8(1 - p) = -12p + 8$

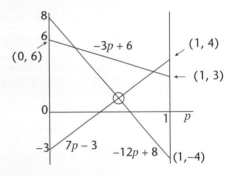

Optimal when $7p - 3 = -12p + 8$

$\Rightarrow p = \frac{11}{19}$

Richard's optimal strategy is to use strategy A with probability $\frac{11}{19}$ and strategy B with probability $\frac{8}{19}$.

Decision Mathematics 2 (continued)

3 (a)

(b)

(c)

The alternating path is:

D – Wed – B – Mon – A – Tue.

The new matching is:

4

Stage	State	Action	Value	
1	G	GT	19	*
	H	HT	14	*
2	D	DG	Max(9, 19) = 19	
	E	EH	Max(17, 14) = 17	*
	F	FH	Max(21,14) = 21	
3	A	AD	Max(12, 19) = 19	*
	B	BE	Max(25, 17) = 25	
		BF	Max(16, 21) = 21	*
	C	CD	Max(23, 19) = 23	
		CF	Max(22, 21) = 22	*
4	S	SA	Max(15, 19) = 19	*
		SB	Max(10, 21) = 21	
		SC	Max(20, 22) = 22	

The optimal minimax route is SADGT.

The maximum distance between supply points on the route is 19 km.

Index